Pavel Em

# CITIES AND THE URBAN SETTLEMENT SYSTEM OF THE REPUBLIC OF KOREA

Pavel Em

# CITIES AND THE URBAN SETTLEMENT SYSTEM OF THE REPUBLIC OF KOREA

## ECONOMIC AND GEOGRAPHICAL ASPECTS OF DEVELOPMENT

**ScienciaScripts**

**Imprint**
Any brand names and product names mentioned in this book are subject to trademark, brand or patent protection and are trademarks or registered trademarks of their respective holders. The use of brand names, product names, common names, trade names, product descriptions etc. even without a particular marking in this work is in no way to be construed to mean that such names may be regarded as unrestricted in respect of trademark and brand protection legislation and could thus be used by anyone.

Cover image: www.ingimage.com

This book is a translation from the original published under ISBN 978-3-8433-1955-3.

Publisher:
Sciencia Scripts
is a trademark of
Dodo Books Indian Ocean Ltd., member of the OmniScriptum S.R.L Publishing group
str. A.Russo 15, of. 61, Chisinau-2068, Republic of Moldova Europe
Printed at: see last page
ISBN: 978-620-3-48741-1

## Contents



**Introduction**

The Republic of Korea has achieved unique economic growth and today ranks 12th in the world in terms of GDP. At the same time, per capita GDP levels are comparable with those of smaller European advanced economies.

The **relevance of the study is** due to the significant political and socio-economic developments in the country which have had a significant impact on its development. In addition, the Republic of Korea belongs to the group of poorly studied post-industrial NIS with an "explosive" urbanisation pattern.

**The object** of the study is the cities and urban settlement system of the Republic of Korea. **The aim of the study** is to identify the economic and geographical aspects of urban development and the urban settlement system of the Republic of Korea. For this purpose it is necessary to solve the following **tasks**:

1)   Conduct a geographical study of urbanisation and a typology of the country's

provinces according to the time of urban transition;

2)        assess the EGPs of the largest cities;

3)  Study the underlying framework of settlement, agglomeration processes and labour integration in agglomerations;

4)  Identify and explain the reasons for the spatial and temporal evolution of the settlement system, and forecast its development.

**The scientific novelty lies in the** fact that, for the first time, a detailed assessment of the hierarchy of cities and urban agglomerations based on the synthesis of classical and relativistic central place theory has been made for the urban settlement system of the Republic of Korea.

The **practical significance** of the study lies in the identification of cities with maximum potential for development, using the methodology developed by the author to assess the EGP of cities, in order to implement plans to relocate the capital of the country. In addition, a forecast of the evolution of the settlement system of the Republic of Korea and unified Korea until 2030 has been made.

This paper has been written using research materials of Russian and foreign scientists. The main statistical basis is the Korean Statistical Yearbook, the Korea Statistical Information Service (KOSIS) and the 2009 World Urban Outlook Report (WUP) [66, 67, 68, 71].

This research was supported by research scholarships for students of Far Eastern State University in 2008 and 2009. Its main points have been published in 12 scientific publications.

The author is grateful to L.I. Ryabinina, Associate Professor at the Department of Regional Analysis and Sustainable Development, Far Eastern State University (Vladivostok), as well as V.A. Shuper, Leading Researcher at the Department of Social and Economic Geography, Institute of Geography, Russian Academy of Sciences, for advice on the research topic and all kinds of help and support.

# Chapter 1: Cities and urban systems as objects of economic and geographical research

## 1.1 Theoretical foundations for the study of cities and urban systems

In contemporary geo-urbanism, paradoxically, there is no unambiguous definition of the term 'city'. Different researchers give different interpretations of the term. For example, E.N. Percik says that a city is "one of the greatest and most complex human creations... the main arena of political, economic and social processes. the place where the greatest values created by man are concentrated" [37]. G.N. Ozerova and V.V. Pokshishevsky [36] refers to a city as "a settlement that performs industrial, organizational and economic, cultural and transport functions". For G.M. Lappo, "the city is a form of settlement and territorial organization of the economy, which has many advantages necessary for social development. The city integrates, melts, transforms, absorbs everything, and it has everything, which is exactly what makes it difficult to define it as a phenomenon" [31]. All cities are multifunctional, dynamic, historically layered, contradictory and problematic. They develop and function in close interaction with the territory around them, serving its needs and finding in it diverse supplements and resources for their development.

"System" is the original philosophical concept of the organisation of the material world as a structurally ordered entity. As it has been introduced into science, the concept of "system" has been refined and clarified. Today there are more than 40 definitions for this concept. The most concise and clear in our opinion is the definition by Alekseev, who states that "a system is a complex of interacting elements" [2]. In geographical science, the systems approach is an adequate research approach in the study of not any objects, but only those of them that represent an organic whole [51].

Any system has the following qualities: *integrity* - the presence of a common purpose and function for the whole combination, which was not the case with a single constituent element; *autonomy* - the desire for greater internal ordering, replenishment of 'missing' elements and functions; *sustainability* - the desire to preserve or develop the structure in such a way that ensures the performance of generalised functions by the system.

An object or a set of objects that perform the same function in a system is called an "element of the system". The main element of a territorial system, which determines its structure and properties, is the settlement system, which is complex, open and complex.

The main theoretical ideas about the settlement systems, the laws of their location and development were developed in the 20th century by the foreign school of economic geography. In 1913, the German scientist F. Auerbach defined the relationship between the size of a city and its rank in the settlement system. According to this law, known as the *"rank-size* rule" (or Zipf, after the man who rediscovered it), the population size of each city in the system is inversely proportional to its rank number in the system:

$$Pr = P / R \ (1),$$

where *Pr* is the population of a given city (rank), *P* is the population of the largest city in the system, *R* is the city rank [39]. The existence of this relationship has been confirmed by numerous scientists who have studied different territories [14, 22, 23, 32, 38, 40, 51, 53]. However, the rule has different deviations in each case study.

In order to find out the advantages of one or another variant of population allocation by settlement centres, A.A. Vazhenin built a simple mathematical model [11]. He presented a matrix in the form of nine cells distributed three by three in width and height. If the cells are considered as settlements, any number of variants can be proposed for the distribution of the population over them. At the same time it is necessary to assume that within one cell as one locality all inhabitants get approximately equal access to satisfaction of their needs. In this case it is interesting to have a relatively regular distribution of the population across the cells. Three options have been proposed: a) absolutely uniform; b) one largest centre and the other eight, divided into four by two levels, decreasing in equal proportion; c) a distribution that satisfies the rank-size rule. In order to be able to obtain comparable figures, the following rules must be set:

1) the number of central functions is directly proportional to the size of the population accommodated in the cell;

2) within their cell, residents of any central location must meet at least 75% of their available needs;

3) one quarter of the functions can be counted when moving to an adjacent cell, and no more than 5% when moving by two cells;

4)       diagonal movement is excluded.

Table 1

Provision of central functions for different settlement cases  (based on [6])

| Cell, no. | Principle of population distribution by cell | | | | | | Estimated number of available central functions | | | | | |
|---|---|---|---|---|---|---|---|---|---|---|---|---|
| | 1 | 2 | 3 | 4 | 5 | 6 | 1 | 2 | 3 | 4 | 5 | 6 |
| 1 | 20 | 45 | 63,627 | 63,627 | 63,627 | 63,627 | 20 | 45 | 63,627 | 63,627 | 63,627 | 63,627 |
| 2 | 20 | 22,5 | 31,814 | 14,547 | 21,695 | 26,09 | 20 | 28,125 | 39,767 | 26,817 | 32,178 | 35,474 |
| 3 | 20 | 22,5 | 21,209 | 14,547 | 21,695 | 26,09 | 20 | 28,125 | 31,814 | 26,817 | 32,178 | 35,474 |
| 4 | 20 | 22,5 | 15,907 | 14,547 | 21,695 | 10,699 | 20 | 28,125 | 27,837 | 26,817 | 32,178 | 23,931 |
| 5 | 20 | 22,5 | 12,725 | 14,547 | 21,695 | 10,699 | 20 | 28,125 | 25,451 | 26,817 | 32,178 | 23,931 |
| 6 | 20 | 11,25 | 10,625 | 14,547 | 7,398 | 10,699 | 20 | 15,188 | 15,377 | 17,001 | 13,069 | 16,424 |
| 7 | 20 | 11,25 | 9,09 | 14,547 | 7,398 | 10,699 | 20 | 15,188 | 14,241 | 17,001 | 13,069 | 16,424 |
| 8 | 20 | 11,25 | 7,963 | 14,547 | 7,398 | 10,699 | 20 | 15,188 | 15,509 | 17,001 | 13,069 | 16,424 |
| 9 | 20 | 11,25 | 7,07 | 14,547 | 7,398 | 10,699 | 20 | 15,188 | 14,847 | 17,001 | 13,069 | 16,424 |
| Total | 180 | 180 | 180 | 180 | 180 | 180 | 180 | 218,25 | 248,47 | 238,9 | 244,61 | 248,13 |
| Average value | | | | | | | 20 | 24,25 | 27,61 | 26,54 | 27,18 | 27,57 |

*Note.* The numbers denote the following columns: **1** - uniform, **2** - decreasing in two steps multiples of 2, **3** - "rank-size", **4** - uniform except the main centre, **5** - decreasing in two steps multiples of 2.9, **6** - crystaller hierarchy K = 3.

By placing a different number of inhabitants in each cell, it is possible to calculate what the total level of central function availability is for the total population of the system. Placing 20 notional residents in each of the 9 cells in a uniform distribution, the resulting sum 180 is distributed across the cells in the other two cases. After calculating the total and average rates of providing central functions to the entire population for each case, it was found that of the proposed population allocation options - the best result gives the "rank-size" allocation (Table 1). The closest values have the Cristaller hierarchy of central location systems (Table 1), which is also very revealing.

The emergence of central place theory was a significant contribution to the formation of modern economic geography. The German scientist Walter Kristaller, studying settlement in southern Germany, found a significant uniformity and hierarchical co-location of all his settlements (65). By examining some indicators of supply and demand for services, he obtained a structure of many hexagons, at the centres of which were the settlements he assigned to different levels of the hierarchy. Comparing the real network of cities and the ideal geometric model, he concluded that there is a certain mechanism that influences the development of settlement systems and leads to the most rational structure [10, 14, 51].

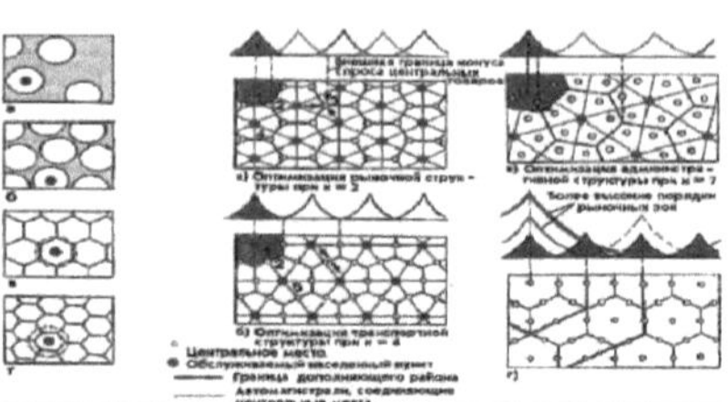

*Figure 1.* Modifications to W. Kristaller's centre seat systems (from [51])

In doing so, while studying the work "The Central Places of Southern Germany" [65], the researchers introduced the necessary conditions for the successful application of Kristaller's theory. **The first postulate** states that space is isotropic in all respects except one - the distribution of urban population, which the theory is intended to describe.

**The second postulate** is the position on the infinity of space. A fragment cannot be arbitrarily extracted from the crystal lattice, because edge effects will occur, which the classical theory can neither describe nor explain.

**The third postulate** is that the zones are as compact as possible (Figure 1). According to it, the system of central places forms a hexagonal lattice of regular hexagons - the most compact and closest to the circle geometric shapes, allowing the most dense packing of settlements in the drukhmer space.

**The fourth postulate** of the principle of optimisation conditions the polymorphism of the central place systems. For example, an optimised market structure leads to a modification with $K = 3$, a transport structure with $K = 4$, and an administrative structure with $K = 7$ (Figure 1). Many researchers define $K$ itself as "the number of lower-level central places of the hierarchy subordinated to one central place, increased by one" (58).

**The fifth postulate** is the assumption of 'rational' consumer behaviour. It assumes that all goods and services are purchased at the nearest central location. However, it is to some extent 'unreliable' because when consumers travel to centres at a certain level of the hierarchy to purchase goods and services, they also purchase what they might find closer to where they live. But since the trip has already been taken, they will use a higher level centre.

Finally, the final **sixth postulate** assumes consistency in the share of the central place in the population of the area it serves for all levels of the crystalline hierarchy [58].

In 1954, the German economist A. Lesh's work "Spacial location of the economy" [32] was published, which presented a more complex model of the location of cities, closer to the real world. He introduced additional factors, the most important of which was the central place common to all settlements of the territory - the largest city, the economic centre of the whole system of settlements. Lesh gave the rationale for the existence of a hexagonal grid and indicated the size of zones and distances between adjacent central locations for different modifications of central locations:

$$A = M - \frac{|AB|}{KK} \qquad (2),$$

where $A$ *is the* distance between adjacent central places; $AB$ *is the* distance between cities of the same hierarchy level [32]. The logic for further reasoning was as follows: the distribution grids with different modifications of $K$ were rotated around a common central place in order to achieve the maximum possible number of centres matching at $K = 3$, $K = 4$, and $K = 7$. This allowed all three structures to be optimised simultaneously. The range of settlement systems considered by the classical central place theory proved to be very limited, and it itself contains many paradoxes leading to the deformation of the hexagonal lattice. These circumstances called for a new approach. In the 1980s, W.A. Schuper proposed a relativistic central place theory that made it possible to overcome some of the contradictions [58]. It made it possible, on the basis of the existing dependence between settlement and space, to explain and mathematically describe settlement systems.

The first problem that the modified Schuper's theory allowed to explain is the explanation in terms of the theory of central places of formation and vitality of urban agglomerations. Crystalier's ideal lattice does not allow for any densification, but agglomerations are places of concentration of urban settlement network. The formation of large agglomerations disturbs not only the predicted proportions in the distance between central places, but also the proportions in the ratios of the sizes of central places at different levels of the hierarchy. The densification of the network of central places near the main centre is compensated by its rarefaction at the periphery, so that the average values for the system as a whole will be close to the case of a regular grid. However, the benefits of concentration accrue both to lower-level central locations because they can use a number of services in it that they would otherwise have to host, and to the main centre, which can use the resources of lower-level central locations to carry out its functions. Harnessing the potential of multiple satellites is an important source of development for a large city without requiring the growth of its size. Due to the excessive concentration of settlements around large cities that occurs in the processes of agglomeration, another phenomenon is observed - the dropout in the settlement system of cities that were supposed to make up the second level of the hierarchy in the region. In this case, the functions of the central places of the second level are taken over by the main city: a curvature of the space of the ideal crystal lattice occurs, in which the cities of the second level of the hierarchy are combined with the main centre into a single agglomeration.

In Schuper's modification, the central place theory has acquired some important indices which enable it to be evaluated from different angles. The most significant are the indicators *of empirical* and *theoretical radii.* The former refers to the real ratio of the distances from the main centre to all the centres of the corresponding hierarchy level and the theoretical values of these distances, found for an ideal lattice as a fraction of the line segment from the centre to the hexagon boundary [58]. The theoretical radius is calculated by the formula:

$$R_{itj} = \frac{\frac{P_{ie}}{P_{it}} * A_i + \frac{P_{je}}{P_{jt}} * A_j}{A_i + A_j} \quad (3),$$

where $P_{ie}$, $P_{je}$, $P_{it}$, $P_{jt}$ are the empirical and theoretical values of cities of different levels; $A_i$, $A_j$ are the proportional shares of responsibility in mutual distances between the levels for a given $K$ [10].

The theoretical population size is determined using the Beckman-Parr equation (formula 4) with a coefficient to determine the hypothetical size of the main city (formula 5):

$$P_h = \frac{P}{h} \quad (4),$$

$$h = \quad (5),$$

where $P_h$ *is* the hypothetical value of the main city, *h* ***is the*** correction factor; $P_1$, $P_2$ *are* the average empirical values of cities of the I and II levels of the hierarchy [10].

The central term of the relativistic modification of central place theory is the notion of *isostatic equilibrium*, which allows a functional relationship to be established between the nature of the spatial organization of urban settlement and the distribution of population between different hierarchical levels. Hierarchical levels of central places are divided into *light* and *heavy*, depending on whether they have a population above or below the theoretically predicted one (formula 4). They usually alternate, thereby 'isostatically balancing' each other. The indicator characterising the state of isostatic equilibrium is the *sum of the ratios of the theoretical radii to the empirical radii* for all highlighted levels of the hierarchy:

$$\sum \frac{R_t}{R} = m - 1 \quad (6),$$

*Rn*

where *m is* the number of hierarchical levels in the system of central places, *c is* the number of missing hierarchical levels [58]. The degree of proximity to the state of isostatic equilibrium can be interpreted as a measure of the stability of he spatial structure of urban settlement systems [10]. The polymorphism of systems consists in the possibility of their existence in various modifications. If the classical theory of Kristaller allowed only variants with $K = 3$, $K = 4$ and $K = 7$, then A.A. Vazhenin [10, 11] and I.A. Khudyaev [56] proved the possibility of the existence of transitional forms with $K = 5$ and $K = 6$ as well as the initial stages of development of systems with $K = 1$ and $K = 2$. Then the whole picture of settlement system evolution takes a complete integrity in the sequential passing of the cycle from $K = 1$ up to $K = 7$ in the process of growing of urban population share. There is a transition region between classical and relativistic theory of central places because neither of them exists in nature in a "pure form". Between these poles lies a vast space to which the vast majority of existing and current urban settlement systems belong [58].

## 1.2 Specifics of urban development and urban settlement systems in new post-industrial economies

The newly industrialised countries (NIS) of the first wave are often referred to as "*Asian tigers*" or "*Asian dragons*". This group includes: Singapore, Hong Kong, Taiwan and the Republic of Korea. These countries have made a qualitative leap in social and economic development in a short period of time. Back in the middle of the last century, they were still industrially undeveloped, predominantly agricultural. Thanks to their rapid development, they now dominate the tertiary sector: 57.6% in the Republic of Korea, 76.2% in Singapore, 92.3% in Hong Kong [70]. The Republic of Korea is of the greatest interest to study the settlement system, due to the small size of the other countries, as well as the uncertain political status of Taiwan. In this country, as in other first-wave NIS, the development of the national economy was oriented towards the foreign market, which was facilitated by TNCs [28]. An example of MNCs in the Republic of Korea are the so-called "*chaebol*" - business conglomerates, the largest of which are world-famous brands: Samsung, LG, Hyundai, Lotte.

The Dragons' distinctive feature is the recent acquisition of independence and progressive development practically "from scratch". For example, the territorial structure of the Republic of Korea's economy began to acquire modern features only after the end of the Korean Civil War in 1953, which almost completely destroyed the country. Almost all of the country's settlements were destroyed along with the economy. The capital city of Seoul, which is located in the immediate vicinity of the demarcation line, suffered particular damage. Busan was relatively 'preserved', which, after the development of the capital began, conditioned the development of a poly-structural settlement system with $K = 2$. The settlement systems of the other Asian Tigers after independence had a structure with $K = 1$, i.e., only one city, the central place, stood out clearly.Due to industrialisation and the rapid development of the NIS, the process of urbanisation proceeded at a very rapid pace. Within 10-30 years of independence, rural countries became urbanised. The second and third phases of urbanisation progressed very rapidly, followed by a marked decrease in the speed of urbanisation as they entered the fourth phase. The dynamic process of urbanization has a paramount influence on the development of central place systems. A.A. Vazhenin identified the threshold values of the share of urban population ($p$), at which the transformation of settlement systems to a higher level takes place [11]. According to his theory, the following levels of urbanization should correspond to a certain structure of the settlement system: at $p$ = 10%, $K = 2$; $p = 30\%$, $K = 3$; $p = 50\%$, $K = 4$; $p = 70\%$, $K = 5$; $p = 90\%$, $K = 6$; $p = 100\%$, $K = 7$. However, this assumption is not supported by the example of the Asian Dragons. For Singapore and Hong Kong, the system with $K = 1$ remains unchanged. In Taiwan and the Republic of Korea, despite more than 75% of the urban population, the settlement system has a modification with $K = 3$. This is due to the particular path of socio-economic development of these countries. All central places are distributed according to the levels of the crystalline hierarchy according to their population size in the proportion 1 - 0.33 - 0.1 - 0.03 [10]. The ratios were maintained in the Republic of Korea, with empirical values corresponding to the theoretical ones with a variation of no more than 5%. In the settlement system of the Republic of Korea, with intensive agglomeration processes, W.A. Schuper's hypothesis of the dropout of the second level of the hierarchy was confirmed. However, while Schuper was able to describe such a phenomenon only on the example of individual regions within states, our study was able to prove the absence of second-tier cities in the country as a whole. This is confirmed in chapter 4 by an analysis of the empirical material. There is a progressive increase in the average size of all levels of the hierarchy, as well as a decrease in the distances between settlements at different levels of the hierarchy. The hexagonal lattice, as the settlement system evolves, becomes progressively more complex due to the emergence of new central locations on it. However, with a considerable number of urban agglomerations and the development of suburbanisation processes that should serve as a mechanism for transforming central place systems, the Republic of Korea's settlement system is "frozen" at a $K = 3$ modification.

**Chapter 2: Urbanisation in the Republic of Korea:  Stages and regional developments**

**2.1 Characteristics and phases of the urbanisation process**

The Republic of Korea has achieved unique economic growth and today ranks 12th in the world in terms of GDP. Only 35 years ago its GDP per capita was comparable to that of the poorest countries in Asia and Africa, but today it is seven times higher than that of India, 13 times higher than that of North Korea and comparable to that of small developed European economies. Over the past 25 years, the annual economic growth rate has been about 8% and in 1986-1988 it was 12% on average [26,28]. - In the past 25 years, annual economic growth has been about 8%, and between 1986 and 1988 it averaged 12% (26, 28).

The progressive modernisation of South Korea's economy in the 1960s and 2005s was accompanied by a rapid increase in urbanisation. A comparison of the country's urbanisation rate with that of one of the world's leading powers, the United States, confirms this. While it took the US 105 years (1890-1995) to go from 40% to 75% urbanisation, it took the Republic of Korea only 20 years (1970-1990) (71). Intensive urbanization, which transformed a "rural" country into an "urban" one, is characterized by temporal and spatial unevenness due to the influence of various economic, social, demographic and external factors.

Urbanization is a multifaceted and multi-component process, the structure of which distinguishes quantitative and qualitative components [43]. The study has considered only the quantitative side of the process, expressed in the increase in the number and proportion of the urban population. Qualitative aspects of urbanization, as well as its social component, were left out of the analysis. This approach is justified by the fact that most countries in East and South-East Asia, with the exception of Japan, Singapore, the Philippines, DPRK and Brunei, have not yet made the urban transition, which, due to weak demographic potential and finding urban development processes in the initial stages of evolutionary development, is not expected in the coming decades [71].

Statistics show that the proportion of the urban population in the Republic of Korea has steadily increased since the formation of the sovereign state in 1948, and continues to the present day (Figure 2). At the same time, the rate of urbanization increased, exceeding 3% from 1960 to 1980. While the proportion of the urban population in the south of the Korean peninsula had been only 21.4% at the time of independence, in 2010 it had risen to 81.9%, an increase of more than 3.8 times (71). Between 1975 and 1980, an important event occurred: the country's urbanization rate exceeded 50%, i.e. the country became "urban" (62).

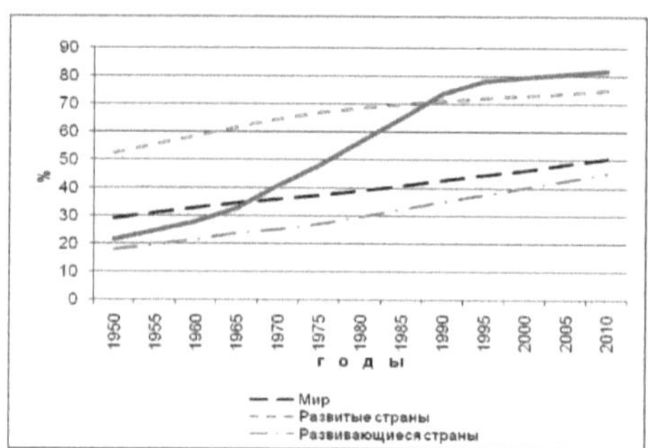



*Figure 2.* Urbanisation in the Republic of Korea against the global backdrop (from [71].)

It is interesting to compare the Republic of Korea against the background of global urbanisation (Figure 2). In 1950, its urban population (21.4%) was lower than the world average (29.1%) and the average for developed countries (52.5%) but higher than that of developing countries (18%). Between 1965 and 1970, the country, through its rapidly growing urban population, surpassed the global average (Figure 2). In 1990, the urban boom which followed the catch-up pattern of lower baseline urbanization and faster rate of urbanization outpaced the developed countries' urban population by 20 years later (Fig. 2). One of the characteristic features of urbanisation is the high growth rate of the large urban population. From 1950 to 1985, it exceeded 5% (Figure 3). In general, the growth rate of the large urban population is always higher than that of the urban population, which means that the bulk of it increased at the expense of large cities. The places of concentration of large-city settlement have created a profound differentiation of the socio-economic space of the country.

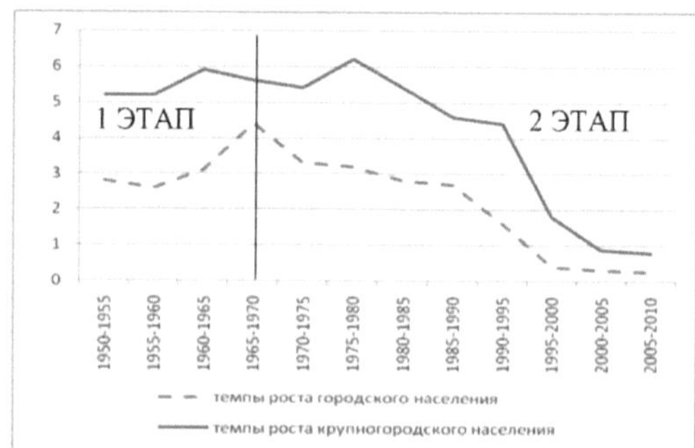



*Figure 3.* Urban and large urban population growth rates in the Republic of Korea (from

13

[71].)

Analysis of the dynamics of the growth rate of the share of the country's urban population during the study period allows us to distinguish between different stages in the intensity of urbanisation (Figure 3).

*The first phase* (1950-1970) witnessed *an accelerated growth of urbanisation* caused by the uneven inflow of investment in provincial infrastructure and its concentration in Seoul. This stimulated the hypertrophied growth of the capital city and the lagging behind of provincial cities in the emerging settlement system. Consequently, the urban population growth rate increased from 2.66% in 1950 to 4.59% in 1970. (Figure 3).

*The second phase* (1971-2010) was characterised by *a slower growth of urbanisation* due to the gradual elimination of provincial inequalities in the distribution of investment, as well as the successful implementation of urban policies under the Urban Planning Act adopted in 1990. [26]. Since 1971, there has been a gradual decrease in the rate of urbanisation. While it was 3.31% from 1970 to 1975, from 2005 to 2010 it did not exceed 0.28% (Figure 3). As a result, the country began a process of urban decentralisation and industrial deconcentration through the relocation of industries from agglomerations to suburbs and satellite towns, and then further across the country.

The number of large cities with more than 100,000 inhabitants has increased considerably over the period under study. If in 1950 there were only 7 of them, in 2009 there were 73. - already 73. At the same time, the number of millionaire cities has increased over the same period from one in 1950 to eight in 2009. The proportion of urban population concentrated in these cities rose from 42% to 64.4%. It is also worth noting the decrease in the degree of hypertrophy of Seoul. While in 1950 its size was 42% of the urban population, in 2009 it was only 27.4%. This is primarily due to the increase in the number of urban settlements, and hence the absolute number of urban dwellers in the country.

Today, two highly urbanised areas have formed in the Republic of Korea: the Seoul-Incheon axis and the Busan-Daegu-Ulsan system, which are the main economic centres of the country and which concentrate about 70% of the country's population in their sphere of influence (63).

Because of the rapid development of urbanisation processes, it is necessary to consider the evolution of this process, because a comprehensive country study requires the study of the relationship between urban and rural populations. Despite the importance of the study of rural settlement, it is the cities, being the main links of the supporting framework of settlement, that determine the territorial structure of the economy and society.

At the beginning of the second half of the twentieth century, scientists drew attention to the universality of urbanization stages, which in each particular country proceed at a different speed and time of occurrence. The first and most optimal theory of spatial evolution, based on the effect of pulsation and cyclical development of settlement, was proposed by J. Jibbs in 1963. It allows us to analyze the process of urbanization based on the minimum amount of evidence - the dynamics of urban and rural population growth rates [21, 29].

Jibbs' periodization provides a comprehensive characterization of urbanization. Taking into account the staging parameters highlighted by the graph of urban and rural population growth rates (Figure 4), the following stages in the evolution of urbanization in the Republic of Korea can be identified.

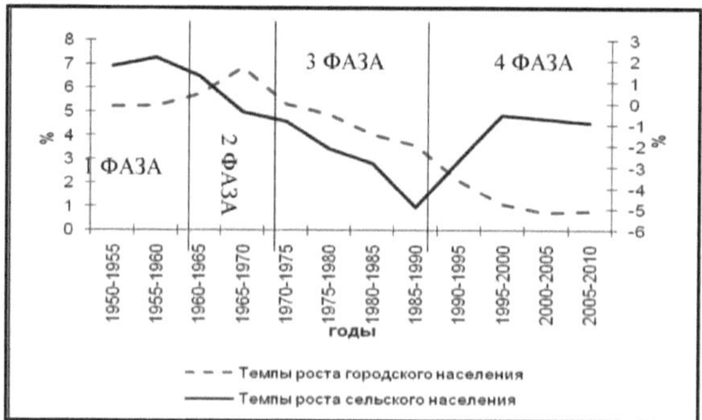


*Figure 4.* Urban and rural population growth rates in the Republic Korea in the second half of the twentieth century. (from The World Urbanization..., 2009)

**The first phase** of urbanization in the south of the Korean peninsula began in the fourth century, when the first cities of Geumseong, Unjin, Sabi, etc. began to be established. This phase was the longest and did not end until 1965. [50]. It was characterized by a predominance of rural inhabitants. In 1965, urban residents made up 32.4% of the country's total population. The absolute growth of the rural population reached 7%, almost 2% ahead of the urban population (figure 4).

The country was characterised by a highly differentiated spatial structure of the economy and population, with the main centres of settlement and economy located on the coast and along the major rivers. The main form of economic activity was agriculture, with a predominance of rice cultivation. Towns, of which there were no more than 15 at the end of the period, had commercial and administrative functions. Almost all of the pre-war cities served as capitals for the new dynasties. Industrialization during this period was at a primitive stage [50].

**The second phase** lasted from 1965 to 1975. Its main feature was that the urban population grew at a higher rate than the rural population (on average 6 and 4% respectively). However, there were still more rural inhabitants, whose share was 59.3%, than urban dwellers [71].

Thanks to a powerful industrial push that economists have dubbed the "miracle on the Hangang River" [28], cities developed as centres of manpower gravity. The lack of job opportunities and hunger, and the lack of jobs and hunger in the cities, have also led to the development of urban settlements. The lack of places of employment and hunger led to an intensification of the labour pendulum migration, as well as to an increase in the centripetal flows of peasants.

Simultaneously with the emerging trends towards concentration of the population, a polarization of the territorial structure of the economy, which has become colonial in nature since the Japanese occupation, is also taking place (63). The first industrial districts are being formed, and the two major industrial hubs of Seoul and Busan, where the main economic activities are concentrated, are being established.

From 1975 to 1990 there is a **third** "turning point" **phase, because the** *urban transition* took place in 1977 (62). The proportion of the urban population rose from 48% to 73.8%. Its growth rate is slightly lower than in the previous period, varying from 4.87 to 5.31% (Figure 4). The high indicators are caused by two factors - a rather high rate of population growth (about 1.5% per year), as well as active migration of rural residents to cities.

The main centres of industrialisation - Seoul, Busan, Ulsan, Pohang, etc. - became the main 'magnets' of the labour force. The differentiation of the socio-economic space of the country has led to the polarisation of areas with the centres and the periphery being singled out. Pendular migration and centripetal flows of rural migration intensified in the latter [63].

**The fourth phase**, which scholars call the "urban revolution", began in 1990 and continues to the present day. It is noteworthy that the proportion of rural inhabitants fell from 26.2% in 1990 to 18.2% in 2010. [71]. Today, there is no longer a rapid increase in the urban population with a parallel decrease in the rural and urban population in small towns (Figure 4). Large pockets of post-industrial development, represented by highly urbanised areas, have emerged in the country. This is facilitated by the rapidly developing 'old' and the large number of new industrial hubs.

With the significant growth of large urban settlements, agglomeration growth has accelerated markedly. Suburbanization processes are becoming more widespread, with increased labour interactions between agglomeration cores and peripheries. The process of suburbanization in the Republic of Korea has developed predominantly based on transport.

Since 1995, the share of the urban population has stabilised. Its growth rate is less than 1%. However, there is as yet no basis for identifying a fifth phase of urbanisation evolution, because the study country is not marked by an active depopulation of rural areas with a parallel development of counter-urbanisation. The transition to the final phase will take place approximately in the mid-1930s, when the share of rural population will be less than 15%, and the currently disrupted rural demography due to female migration in 1970-1980 will lead to a true depopulation of the village. In 2050, UN experts predict that the proportion of the rural population in the Republic of Korea will not exceed 10.2% [71].

The main feature of the evolution of urbanisation in the Republic of Korea is the prolonged onset and very rapid period change due to the 'explosive' nature of the process.

## 2.2 Urban typology of the country's provinces

What is significant about the nature of urbanisation in the South Korean provinces is that they became "urban" at about the same time as the country (1970-1975) and have a similar pattern of urban population share and growth. The countrywide trend is universal and is repeated in most provinces with minor variations, so it can be called "standard" (Figure 5).

To describe spatial differences in the urbanization of the country, we conduct a typology of its provinces based on the methodological approaches of R.A. Popov [44] and O.A. Konstantinov [24], in which *the share of urban population is* taken as a static result of urbanization. It is proposed to distinguish the following periods of urban transition of the provinces: 1) before 1955; 2) 1956-1975; 3) 1976-1985. Each of the periods is assigned a specific urbanisation type.

Each type can be characterised by the ranges of initial (as of 1975) and growth rates for the urban population calculated as the ratio of the final value (as of 2005) to the initial value (Figure 5), common to all provinces. Varying these parameters within the ranges allows us to divide the second type of provinces into sub-types (Table 2).

Table 2 Typology of provinces in the Republic of Korea according to urbanisation levels from 1975 to 2005.

(compiled by the author on the basis of calculations)

| № | Type | Subtype | Provinces | Features of the sub-type | |
|---|---|---|---|---|---|
| | | | | Proportion of urban population in 1975 (%) | Urban population growth rate (times) |
| 1 | A province with an early urban transition (before 1955) | 1.1 | Gyeonggi-do | Over 55 | 1,5 |
| 2 | Provinces with a relatively early urban transition (1956-1975) | 2.1 | South Chungcheong North Gyeongsang South Gyeongsang South Jeolla | 50-55 | 1,6 |
| | | 2.2 | Gangwondo Jeju-do North Cholla | 40-50 | 1.6 |
| 3 | A province with a medium urban transition time (1976-1985) | 3.1 | Chuncheon North | 35-40 | 1,8 |

Trends in the proportion of urban population in the provinces and their growth rates differ significantly from the "standard" one. The geographical distribution of the identified urbanisation types in the Republic of Korea is shown in Figure 6. 6. Let us now look at them directly.

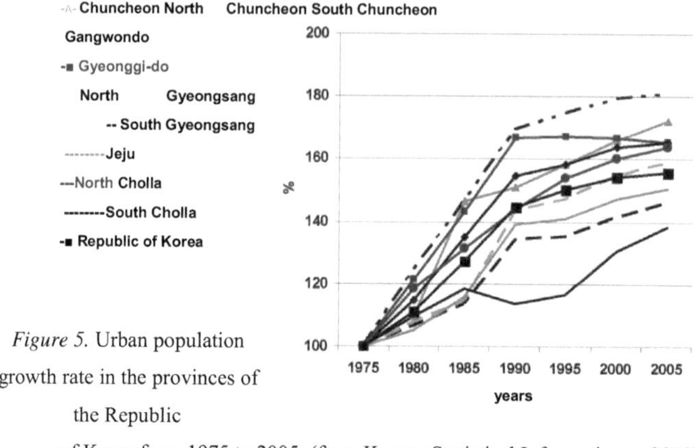


*Figure 5.* Urban population growth rate in the provinces of the Republic of Korea from 1975 to 2005. (from Korean Statistical Information..., 2010)

**Type 1. A province with an early urbanisation transition (before 1955).** The first type is represented by a province that embarked on the path of urbanisation and achieved a high level of urbanisation earlier than others. Here the share of urban population increased rapidly in the second quarter of the 20th century and the urbanisation process took place after the Korean War. By 1950, its urbanisation rate was well on its way to reaching its maximum possible level, exceeding 55%. Accordingly, its growth rate over the period under review is minimal (1.5 times).

Urbanization has developed most rapidly in Gyeonggi-do Metropolitan Province. The proportion of the urban population there was already over 58% in 1975 and exceeded 98% in 1995 (66). Here, the specific 'metropolitan' status of the province, implying such factors as the diversity of workplaces, high level of infrastructure development, investment attractiveness, etc. played a key role. The attraction of the capital has contributed to the urbanization of the province: the first agglomeration in the Republic of Korea has been formed here. Gyeonggi-do province still leads the country in terms of the total number of urban settlements.

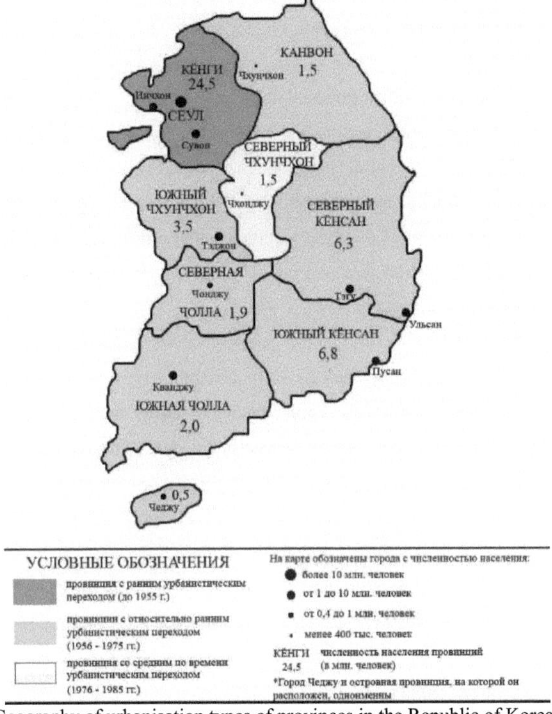


*Figure 6.* Geography of urbanisation types of provinces in the Republic of Korea from 1975 to 2005. (compiled by the author on the basis of calculated data)

Gyeonggi-do crossed the 80% urban population threshold after 1980 through administrative border expansion and the rapid pace of rural migration in Seoul and Incheon.

**Type 2. Provinces with a relatively early urbanisation transition (1956-1975).** The urbanisation trend of this type of province is closest to the national average. Compared to the first type provinces, they had a later urbanisation transition, in the 1975-1985s, and a lower initial urbanisation rate, ranging from 44% to 49%. In most cases, due to a higher growth rate, by the end of the period in question, the rate had risen to a level of 68-80%, i.e. reaching values comparable to those of the first type (66). The urbanisation growth rate for this type of province was 1.6.

*Subtype 2.1* is very representative: it combines one western Korean province, two southern provinces, and one eastern province. They are characterised by a slightly lower initial urbanisation rate than Type 1, ranging from 50 to 55%. The final values, in turn, range from 71% to 92%.

The provinces of North and South Gyeongsang as well as South Jeolla have emerged as urban 'leaders' thanks to early and vigorous industrial development, entailing a growing urban population. Most of the hotbeds of new industrialization are located in their territory. As for South Chungcheong Province, it is highly urbanized primarily because of local settlement patterns. Here the density of rural settlements is low, cities are not numerous, but their average population density is quite high, which is typical of a focal and sparsely focal settlement area.

*Subtype 2.2* is the agricultural province of North Cholla with a long-established and sparse network of urban settlements. Its long-standing lag in urbanisation levels compared to Type 1 has resulted in a higher rural population density and a lower proportion of regional centres. It is a kind of periphery in relation to the more urbanised provinces.

In contrast, Jeju-do has not historically had a well-developed urban network, but its administrative centre has become a major city. Jeju and Seohipyo are the cities that account for the bulk of the province's population. Gangwon-do occupies an intermediate position between the two groups of provinces.

**Type 3: A province with an average urban transition time (1976-1985).** This type is represented by one central province, North Chungcheong, characterized by "average" urbanization parameters. In the early 1975s the proportion of urban population was only 37%, but it has grown by a factor of 1.8 over 30 years and in 2005 it was just over 64% (66). To this day, the proportion of the population living in urban areas in the province has not stabilised, here is still an upward trend. This is a consequence of the colonial type of territorial economic structure and the late start of urbanisation.

## Chapter 3: Cities in the Spatial Structure of the Economy of the Republic of Korea

### 3.1 Analysis of the economic and geographical position (EPG) of cities

#### 3.1.1. Urban EGP assessment methodology

In economic geography, location is defined as "the relationship of a place, an area, or a city to the outlying features of economic importance" [4]. EGP is a specific resource that plays a decisive role in the development of a city [3]. I.M. Maergoyz wrote that "to know a city geographically means to understand its geographical position" [33]. Therefore, this section attempts to synthesize three different ways to help assess the EGP of cities. *The topological approach* made it possible to identify the level of connectivity and systemicity of each city. For this purpose, a connectivity graph was constructed, with cities with a population of more than 0.5 million people as nodes and the main railway lines of the country as lines (Figure 7). On its basis, the adjacency matrix was compiled (Table 4), which made it possible to determine the number of neighbours of any rank for each territorial unit, as well as the presence of neighbours along the shortest path to assess the connectivity of the network in question. For this purpose, three indicators were calculated based on the matrix data:

1 *total neighbourhood index* ($si$)

$$Si = z \overset{n}{NR} \ (7).$$

where $N$ is the total number of system elements under consideration;

2 *the average neighbourhood rank of the element in the network in question* ($R\text{-}J$

$$R_{vo} : .. \quad \frac{Si}{N11} \ (8);$$

3 *average neighbourhood rank in the whole network* (*Keti*)

$$_{networks} \hspace{4cm} (9).$$

*The method of potentials* is effective for investigating theoretically possible relations. It is based on the gravity model, which is applied when it is possible to describe the interaction of two elements as a physical model of gravitational masses. The induced potential formula was used to calculate the potential:

$$_{iv} \sim P \ (10),$$
$$_{..1} d_{,i}$$

where $_{iv}$ is the induced potential at point $j$; $P_f$ is the mass of -- points (population size); $dj$ is the distance from point $j$ to $i$ points (air). The value of the induced potential is taken as a potential measure of the influence on a given point from outside [34].

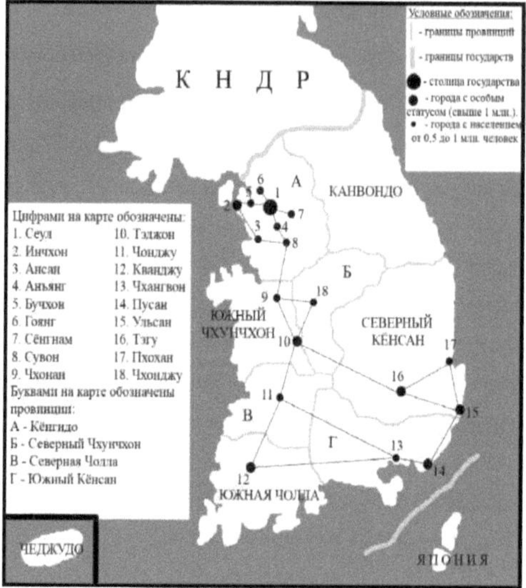



*Figure 7.* Main elements of the settlement framework of the Republic of Korea (compiled by the author)

The study of *transport and geographic location (TGE)* is an integral component of the assessment of the EGP (Topchiev, 1974). The classification of the major cities of the Republic of Korea according to the potential TGP was based on a scoring of the level of formation of transport hubs according to technical characteristics, the degree of development of connections for each mode of transport and the transport complex in general, as well as the degree of use of the maritime position. The analysis of the first two indicators was based on the construction of matrices, the columns of which indicate the degree of formation of the transport complex, and the rows - the degree of connectivity [12]. The matrix columns are divided into 5 blocks corresponding to the following stages of transport development:

4 - the most technically simple transport points (railway platforms, junctions, marinas);

5 - railway stations, river, sea and air ports;

6 - hubs for rail, road, water and air transport of provincial importance;

7 - large transport complexes of national and international importance;

8 - Centres that have developed scientific services and trained highly qualified

transport professionals.

According to the scale of linkages, the matrices are divided into 6 blocks with rows corresponding to the following types of linkages: 1 - intra-city, 2 - local, 3 - provincial, 4 - regional, 5 - state and international. The resulting totals give an indication of *the degree of maturity of the* urban *transport complex (UTC)*.

Each city's position in terms of major modes of transport was analysed by stage of development and the overall score was then determined. Table 3 is an example of the NWTK calculation for Busan. The scores were added up as follows: rail (1+4+6+8+9), road (1+3+5+8=8), water (1+3+5+5+9), and air (5+6+9+9). The total score is 109. The scores for the other cities are shown in Table 5.

Table 3

Example of transport maturity calculation for Busan

(based on [67, 68])

| | Railway | | | | | Automotive | | | | | Water | | | | | Air | | | | |
|---|---|---|---|---|---|---|---|---|---|---|---|---|---|---|---|---|---|---|---|---|
| | 1 | 2 | 3 | 4 | 5 | 1 | 2 | 3 | 4 | 5 | 1 | 2 | 3 | 4 | 5 | 1 | 2 | 3 | 4 | 5 |
| 1 | + | | | | | + | | | | | + | | | | | | | | | |
| 2 | | | | | | | + | | | | | + | | | | | | | | |
| 3 | | + | | | | | | + | | | | | + | | | | | | | |
| 4 | | | + | | | | | | | + | | | | | | + | + | | | |
| 5 | | | | + | + | | | | + | | | | | | + | | | | | + |
| 6 | | | | | | | | | | | | | | + | | | | | + | |

This indicator does not fully reflect the fact that of the 18 large cities of the country 11 are located in the coastal strip up to 50 km, and 7 - from 50 to 200 km and they use the resource of the coastal position differently. Therefore, the study introduces an additional synthetic indicator - *the degree of use of coastal position (SIPP),* obtained as a result of scoring the number of population and the value of urban product (tab. 5) [5].

### 3.1.2. Results of an assessment of the EGP of the country's largest cities

Using the topological approach, total neighbourhood scores (TC) and average neighbourhood rank (SRS) were calculated (Table 4). The higher the EGP potential is, the lower the CPC is. The minimum CPC is characteristic of Cheonan (2.4) and the maximum is characteristic of Goyang and Seongnam (4.3).

Table    4

Summary matrix for shortest distance and average neighbourhood rank

between the largest cities in the Republic of Korea (from Figure 7)

| | 1 | 2 | 3 | 4 | 5 | 6 | 7 | 8 | 9 | 10 | 11 | 12 | 13 | 14 | 15 | 16 | 17 | 18 | CC | SRS | EGP |
|---|---|---|---|---|---|---|---|---|---|---|---|---|---|---|---|---|---|---|---|---|---|
| 1 | - | 2 | 3 | 1 | 1 | 1 | 1 | 2 | 3 | 4 | 5 | 5 | 6 | 7 | 6 | 5 | 6 | 4 | 62 | 3,4 | 1 |
| 2 | 2 | - | 1 | 3 | 1 | 3 | 3 | 2 | 3 | 4 | 5 | 5 | 6 | 7 | 6 | 5 | 6 | 4 | 66 | 3,7 | 1 |
| 3 | 3 | 1 | - | 2 | 2 | 4 | 4 | 1 | 2 | 3 | 4 | 4 | 5 | 6 | 5 | 4 | 5 | 3 | 58 | 3,2 | 2 |
| 4 | 1 | 3 | 2 | - | 2 | 2 | 2 | 1 | 2 | 3 | 4 | 4 | 5 | 6 | 5 | 4 | 5 | 3 | 54 | 3 | 3 |
| 5 | 1 | 1 | 2 | 2 | - | 2 | 2 | 3 | 4 | 5 | 6 | 6 | 7 | 8 | 7 | 6 | 7 | 5 | 74 | 4,1 | 1 |
| 6 | 1 | 3 | 4 | 2 | 2 | - | 2 | 3 | 4 | 5 | 6 | 6 | 7 | 8 | 7 | 6 | 7 | 5 | 78 | 4,3 | 1 |
| 7 | 1 | 3 | 4 | 2 | 2 | 2 | - | 3 | 4 | 5 | 6 | 6 | 7 | 8 | 7 | 6 | 7 | 5 | 78 | 4,3 | 1 |
| 8 | 2 | 2 | 1 | 1 | 3 | 3 | 3 | - | 1 | 2 | 3 | 3 | 4 | 5 | 4 | 3 | 4 | 2 | 46 | 2,6 | 3 |
| 9 | 3 | 3 | 2 | 2 | 4 | 4 | 4 | 1 | - | 1 | 2 | 2 | 3 | 4 | 3 | 2 | 3 | 1 | 44 | 2,4 | 3 |
| 10 | 4 | 4 | 3 | 3 | 5 | 5 | 5 | 2 | 1 | - | 1 | 1 | 2 | 3 | 2 | 1 | 2 | 1 | 45 | 2,5 | 3 |
| 11 | 5 | 5 | 4 | 4 | 6 | 6 | 6 | 3 | 2 | 1 | - | 1 | 1 | 2 | 3 | 2 | 3 | 2 | 56 | 3,1 | 2 |
| 12 | 5 | 5 | 4 | 4 | 6 | 6 | 6 | 3 | 2 | 1 | 1 | - | 1 | 2 | 3 | 2 | 3 | 2 | 56 | 3,1 | 2 |
| 13 | 6 | 6 | 5 | 5 | 7 | 7 | 7 | 4 | 3 | 2 | 1 | 1 | - | 1 | 2 | 3 | 3 | 3 | 66 | 3,7 | 1 |
| 14 | 7 | 7 | 6 | 6 | 8 | 8 | 8 | 5 | 4 | 3 | 2 | 2 | 1 | - | 1 | 2 | 2 | 4 | 76 | 4,2 | 1 |
| 15 | 6 | 6 | 5 | 5 | 7 | 7 | 7 | 4 | 3 | 2 | 3 | 3 | 2 | 1 | - | 1 | 1 | 3 | 66 | 3,7 | 1 |
| 16 | 5 | 5 | 4 | 4 | 6 | 6 | 6 | 3 | 2 | 1 | 2 | 3 | 3 | 2 | 1 | - | 1 | 2 | 56 | 3,1 | 2 |
| 17 | 6 | 6 | 5 | 5 | 7 | 7 | 7 | 4 | 3 | 2 | 3 | 4 | 3 | 2 | 1 | 1 | - | 3 | 69 | 3,8 | 1 |
| 18 | 4 | 4 | 3 | 3 | 5 | 5 | 5 | 2 | 1 | 1 | 2 | 2 | 3 | 4 | 3 | 2 | 3 | - | 52 | 2,9 | 3 |

*Note.* The following cities are marked with numbers: 1 - Seoul, 2 - Incheon, 3 - Ansan, 4 - Anyang, 5 - Bucheon, 6 - Goyang, 7 - Seongnam, 8 - Suwon, 9 - Cheonan, 10 - Daejeon, 11 - Cheongju, 12 - Gwangju, 13 - Changwon, 14 - Busan, 15 - Ulsan, 16 - Daegu, 17 - Pohang, 18 - Cheongju.

It should be pointed out that in the distribution of the quantitative data obtained for the cities, we have designated groups of potential EGP corresponding to the following numbers: 1 - low, 2 - medium, 3 - high. In order to group the cities according to NDN, we have divided the values obtained into 3 equal intervals, the upper and lower limit of which have been set by the maximum and minimum values, and the intermediate ones so that they divide the interval into 3 equal parts. Thus, the cities according to the CDS can be categorised into the following groups: 1 - Seoul, Incheon, Bucheon, Goyang, Seongnam, Busan, Ulsan, Changwon, Pohang; 2 - Ansan, Jeonju, Gwangju, Daegu; 3 - Cheongju, Daejeon, Suwon, Cheonan, Anyang.

The results obtained using the topological method lead to the conclusion that the highest potential EGP is characteristic of cities gravitating towards the geometric centre of the country. This is due to the specificity of the method, as the SRF will be smaller the closer the city is to the centre. This method is very indicative, but it does not take into account all the specifics of the EGP of cities, as it considers only a particular part of it.

*Figure 8.* Assessment of the economic and geographical location of the largest cities of the Republic of Korea (compiled by the author on the basis of calculated data)

Table 5

Assessment of the degree of maturity of the transport complex and the degree of utilisation of the coastal position of the largest cities in the Republic  of Korea (compiled by the author on the basis of calculated data)

| № | City | C3TK EVALUATION | | | | | Score by NWTC | EVALUATING THE CIPP | | | | Score CIPP | Total score NWTC and CIPP | TGP |
|---|---|---|---|---|---|---|---|---|---|---|---|---|---|---|
| | | Scores for individual modes of transport | | | | | | Population, (in thousands) | Score by population | GDP per capita (US$) | Score by GDP | | | |
| | | w/o | Automobile | water | air | Score sztk | | | | | | | | |
| 1 | Busan | 28 | 25 | 27 | 29 | 109 | 3 | 3615,1 | 2 | 19698 | 1 | 3 | 6 | 3 |
| 2 | Incheon | 26 | 23 | 8 | 28 | 85 | 3 | 2710 | 2 | 24103 | 1 | 3 | 6 | 3 |
| 3 | Ulsan | 17 | 17 | 28 | 12 | 74 | 2 | 1112,8 | 1 | 58665 | 3 | 4 | 6 | 3 |
| 4 | Seoul | 27 | 28 | 5 | 7 | 67 | 2 | 10421,8 | 3 | 31095 | 2 | 5 | 7 | 3 |
| 5 | Taejon | 18 | 18 | 0 | 12 | 48 | 2 | 1487,89 | 1 | 29537 | 2 | 3 | 5 | 2 |
| 6 | Gwangju | 17 | 18 | 0 | 12 | 47 | 2 | 1423,5 | 1 | 29412 | 2 | 3 | 5 | 2 |
| 7 | Tag | 17 | 17 | 0 | 12 | 46 | 2 | 2512,7 | 2 | 17217 | 1 | 3 | 5 | 2 |
| 8 | Cheongju | 17 | 17 | 0 | 12 | 46 | 2 | 638,4 | 1 | 15028 | 1 | 2 | 4 | 1 |
| 9 | Pohang | 10 | 10 | 17 | 4 | 41 | 1 | 508,7 | 1 | 20426 | 1 | 2 | 3 | 1 |
| 10 | Cheongju | 17 | 18 | 0 | 4 | 39 | 1 | 627,3 | 1 | 17470 | 1 | 2 | 3 | 1 |
| 11 | Suwon | 18 | 15 | 0 | 4 | 37 | 1 | 1086,9 | 1 | 36345 | 2 | 3 | 4 | 1 |
| 12 | Cheonan | 17 | 15 | 0 | 4 | 36 | 1 | 540,7 | 1 | 21080 | 1 | 2 | 3 | 1 |
| 13 | Goyang | 14 | 14 | 0 | 4 | 32 | 1 | 935,6 | 1 | 37407 | 2 | 3 | 4 | 1 |
| 14 | Ansan | 8 | 16 | 0 | 4 | 28 | 1 | 734,7 | 1 | 46362 | 3 | 4 | 5 | 2 |
| 15 | Changwon | 9 | 9 | 0 | 4 | 22 | 1 | 510,1 | 1 | 26481 | 1 | 2 | 3 | 1 |
| 16 | Bucheon | 8 | 8 | 0 | 4 | 20 | 1 | 876,6 | 1 | 41643 | 2 | 3 | 4 | 1 |
| 17 | Seongnam | 0 | 15 | 0 | 4 | 19 | 1 | 968,2 | 1 | 42314 | 2 | 3 | 4 | 1 |
| 18 | Anjang | 0 | 8 | 0 | 4 | 12 | 1 | 630,7 | 1 | 35257 | 2 | 3 | 4 | 1 |

*Note.* Numbers indicate groups corresponding to: 1 - low, 2 - medium, 3 - high THP.

Through the use of the potential method (formula 10), we have calculated the potentials of the cities induced by the rest of the settlements of the set census (Figure 8, Table 6).

When using this methodology, the opposite rule works: the higher the potential, the higher the EGP. It has been shown to have high potential for most cities. Only six cities were the exceptions.

The group with medium potential includes: Busan, Incheon, Daejeon, Daegu, and Seongnam; and Seoul with low potential. The analysis of the data obtained on EGP potential helps to conclude that the more developed a city is at present, the less development potential it has, and vice versa, the less developed it is, the higher the potential for further development. This point is proved by the example of the study country. The state capital, the most powerful economic centre, has a potential of only 0.17, while 'peripheral' cities like Cheongju and Changwon have more than 0.28.

Table 6 Potentials of the economic and geographical position of cities in the Republic of Korea, calculated by the potential method (compiled by the author using estimated data)

| № | City | Calculated Capacity | Level of EGP | № | City | Calculated capacity | Level of EGP |
|---|---|---|---|---|---|---|---|
| 1 | Busan | 0,24 | 2 | 10 | Cheongju | 0,28 | 3 |
| 2 | Incheon | 0,25 | 2 | 11 | Suwon | 0,27 | 3 |
| 3 | Ulsan | 0,27 | 3 | 12 | Cheonan | 0,28 | 3 |
| 4 | Seoul | 0,17 | 1 | 13 | Goyang | 0,27 | 3 |
| 5 | Taejon | 0,27 | 2 | 14 | Ansan | 0,27 | 3 |
| 6 | Gwangju | 0,27 | 3 | 15 | Changwon | 0,28 | 3 |
| 7 | Tag | 0,25 | 2 | 16 | Bucheon | 0,27 | 3 |
| 8 | Cheongju | 0,28 | 3 | 17 | Seongnam | 0,27 | 2 |
| 9 | Pohang | 0,28 | 3 | 18 | Anjang | 0,28 | 3 |

The results of a comprehensive analysis of the SSTC of the largest cities in the Republic of Kazakhstan indicate the lack of maturity of the transport complex as a whole. It is not the capital, but the largest southeastern port, Busan, that has the highest score for the NTPC (Table 5). This can be explained by the high degree of the city's transport complex with the most extensive range of connections. Busan's leadership in terms of overall NWTK score is undeniable, as Incheon, which has only 85 points, occupies the second rank. The amplitude of the NWTC value is significant and is 107 (with Anyang in 18th place with only 12 points) (Table 5). This proves the excessive unevenness of transport development in different parts of the country. The maximum development of transport thoroughfares has taken place near the metropolitan cities, which once again emphasises their special role in the territorial division of labour. These same cities also have the largest population, ranging from 1.11 to 10.422 million, ranking in the top seven places in the structure of the urban settlement components of the country (Table 5).

The coastal location has a rather strong influence on the level of SSTD of each individual settlement. It is easy to see that the maximum SSTC scores are exclusively characteristic of towns no more than 50 km from the seacoast, i.e. those directly located in the coastal zone. There is a trend here: the more continental a city is, the fewer people live in it (64).

The level of gross urban product per capita is also highly differentiated. The ratio of the maximum and minimum values specific to Ulsan and Daegu ($58665 and $17217) is 3.4 (Korea Statistical Yearbook, 2009). The GRP of "continental" cities "indirectly connected to the sea" is significantly lower than that of cities located within 50 km of the coastal zone and port cities (64).

The most important transportation centres in the country, Busan, Incheon, Seoul, and Ulsan, have been assigned to the group with high TGP potential. The remaining metropolitan cities in the middle group are Daegu, Gwangju, Daejeon, and Ansan due to their very high per capita urban product (over $46,000 per capita). All other cities have weak TPP potential (Table 5).

With the data obtained, we assigned points to each city, using the sum of which we categorised them into three types according to their total potential for EGP (Table 7). It is worth noting that each of the methods has its own advantages and disadvantages. However, their combined use made it possible to carry out a fairly complete assessment of the EGP of the cities with a set price tag.

The first type (with a high potential for EGP) includes 8 cities. Their location in relation to the sea coast is differentiated. In addition to the port city of Ulsan, it includes cities that are "indirectly connected to the sea" and that gravitate toward the geometric centre of the country (Gwangju and Daejeon). The vast majority of non-metropolitan cities are concentrated in the metropolitan area, with the remaining cities in neighboring provinces (in the northern part of the country).

Table 7 Comprehensive assessment of the economic and geographical potential of major cities in the Republic of Korea (author's calculations)

29

| City | EGP 1 | EGP 2 | TGP | Amount EGP and TGP | Type EGP | City | EGP 1 | EGP 2 | TGP | Sum of EGPs | Type EGP and GP |
|---|---|---|---|---|---|---|---|---|---|---|---|
| Busan | 1 | 2 | 3 | 6 | 2 | Cheongju | 2 | 3 | 1 | 6 | 2 |
| Incheon | 1 | 2 | 3 | 6 | 2 | Suwon | 3 | 3 | 1 | 7 | 3 |
| Ulsan | 1 | 3 | 3 | 7 | 3 | Cheonan | 3 | 3 | 1 | 7 | 3 |
| Seoul | 1 | 1 | 3 | 5 | 2 | Goyang | 1 | 3 | 1 | 5 | 2 |
| Taejon | 3 | 2 | 2 | 7 | 3 | Ansan | 2 | 3 | 2 | 7 | 3 |
| Gwangju | 2 | 3 | 2 | 7 | 3 | Changwon | 1 | 3 | 1 | 5 | 2 |
| Tag | 2 | 2 | 2 | 6 | 2 | Bucheon | 1 | 3 | 1 | 5 | 2 |
| Cheongju | 3 | 3 | 1 | 7 | 3 | Seongnam | 1 | 2 | 1 | 4 | 1 |
| Pohang | 1 | 3 | 1 | 5 | 2 | Anjang | 3 | 3 | 1 | 7 | 3 |

*Note.* EGP 1 refers to results obtained using the topological approach (Table 4), EGP 2 using the potentials method (Table 5).

The second type is the most representative, comprising nine cities (including five cities with special status) with moderate EGP potential. It is noteworthy that this group, along with the economically "lagging behind," includes the country's most developed cities: Seoul, Busan, and Incheon. This is due to the theoretical exhaustion of the development potential of these areas. This has been realised by the government of the Republic of Korea, which is trying to pursue an urban policy aimed at decentralising industry and deconcentrating the population [29].

Notably, only Seongnam has a low potential for EGP. As a city directly included within the Seoul agglomeration, it has in fact recently become part of the Greater Seoul area, and is a 'bedroom' appendage of the capital.

**3.2 Assessing the maturity of the settlement support framework (SSF)**

**3.2.1 Nodal elements of the RCD**

All cities in the Republic of Korea were considered as anchor frame nodes, but urban agglomerations with more than 500,000 inhabitants were the main elements of the study. It was necessary to determine their composition based on the assessment of the role played by one or another centre in the territorial organization of society. The main criteria were: combination and scope of functions performed, as well as geographical location. The basis of the set of nodal elements of the support frame of the Republic of Korea is formed by the hierarchically constructed system of cities - central places, supplemented by the largest centres of different industries, major transport hubs.

In terms of the total number of millionaire agglomerations, the Republic of Korea ranks third in the Asia-Pacific Region (APR) after the People's Republic of China (PRC) and Japan (6 agglomerations with a total population of 34.052 million).

The analysis of administrative-territorial localization of nodal elements of the basic framework of settlement of the Republic of Korea by groups of cities in 2009. (Table 8) suggests that the largest number of urban settlements of all groups in the number of 29 units (grouping is made by the method of G.M. Lappo [31]) is concentrated on the territory of the Capital Region. Most of its cities are large cities with a population of more than 250 thousand people; two of them are millionaires. It is worth noting that 3 other cities (Bucheon, Seongnam, and Yeongjin) in Gyeonggi-do Province are approaching million-dollar status and are also projected by UN experts to be of the above size by 2020 [71].

A similar pattern is found in the southeast of the country, where there are 21 urban settlements in North Gyeongsang and South Gyeongsang provinces. Three of them are metropolitan areas and form a closely interconnected system. The two regions considered (the Capital Region and the southeast of the country) concentrate about 70 percent of the country's population in their area of influence [62].

31

Administrative-territorial localisation of the core

settlement pattern nodes

of the Republic of Korea by city grouping

(based on [66])

| ATE / Type cities | Number of cities | | | | | | Total cities |
|---|---|---|---|---|---|---|---|
| | small (up to 50,000 people) | medium-sized (50,000 to 100,000 people) | large (100,000 to 250,000 people) | large (250,000 to 500,000 people) | Largest (0.5 to 1 million people) | millionaire cities | |
| Special purpose cities | 0 | 0 | 0 | 0 | 0 | 7 | 7 |
| Gyeonggi-do | 0 | 2 | 10 | 8 | 6 | 1 | 27 |
| Gangwondo | 0 | 4 | 1 | 2 | 0 | 0 | 7 |
| North Chuncheon | 0 | 0 | 3 | 0 | 1 | 0 | 4 |
| Southern Chuncheon | 1 | 0 | 4 | 0 | 1 | 0 | 6 |
| North Gyeongsang | 0 | 1 | 6 | 0 | 1 | 0 | 8 |
| Southern Gyeongsang | 0 | 0 | 6 | 4 | 1 | 0 | 11 |
| North Cholla | 0 | 2 | 3 | 2 | 1 | 0 | 8 |
| South Cholla | 0 | 1 | 1 | 2 | 0 | 0 | 4 |
| Jeju-do | 0 | 0 | 1 | 1 | 0 | 0 | 2 |
| **TOTAL** | 1 | 10 | 35 | 19 | 11 | 8 | 84 |

In general, the country's settlement system is dominated by
The smallest town is Geryeong, in South Chungcheong Province, with only 38,000 inhabitants
(Korean Statistical Statistics). The smallest in terms of population and the only locality that
belongs to the group of small towns is Geryong in South Chungcheong Province with only
38,000 inhabitants (Korean Statistical Information..., 2010). The major cities (Cheongju,
Cheonan, Pohang, Changwon, and Jeonju) located outside the highly urbanized areas are the

main economic centres of the semi-periphery and periphery and serve as the main elements of the supporting framework of settlement in the centre and southwest.

South Korean cities have a fairly large area, which includes agricultural land. Only 19% of urban land is occupied by buildings, roads, and industries (67).

Based on the level of urbanisation, concentration of population in cities with different population density and settlement patterns, the Republic of Korea is a highly urbanised country with a predominant and fast-growing population in large and major cities (Table 8). Settlement in the country is tied to the plains, so the most densely populated areas are the northwest and southeast of the Republic. In the mountainous eastern and north-eastern areas the population density is significantly lower.

### 3.2.2 Linear elements of ROC

The linear-nodal subsystem of RCD at the present stage is represented not just by transport arteries, but by polimahighways consisting of two or more infrastructure lines. It performs a "nutrient-distributive or circulatory role", ensuring the connection of the most important transport and economic nodes and areas intensively interacting with each other [42]. That is why it is necessary to study the entire transport system, which is one of the key indicators of economic development of the territory, because it is known that without the presence of transport infrastructure of a certain quality level its sustainable development is impossible.

The modernisation of the Republic of Korea's economy in the second half of the twentieth century was the result today [67]:

- The length of railways in operation is 3.12 thousand km;
- The length of paved roads is 91,500 km;
- There are over 13 million vehicles and 350,000 transport companies in the country;
- More than 3% of the economically active population is employed in transport;

- Transport companies account for 7.3% of the country's total GDP structure.

Analysis of the dynamics of cargo and passenger flows in the country has allowed for the identification of 2 stages (Figure 9). *The first stage* (1980-1990) saw a progressive increase in both cargo and passenger volumes. The increase in the former was due to the increase in the volume of export products, which were transported through the main ports of Busan and Ulsan. The increase in passenger traffic is due to the development of civil aviation as well as passenger ferry services in the south of the country.

With the beginning of *the second phase* (1990-2007), there is a decrease in gross passenger traffic. The main reason is the increase in the fleet of private vehicles [60].

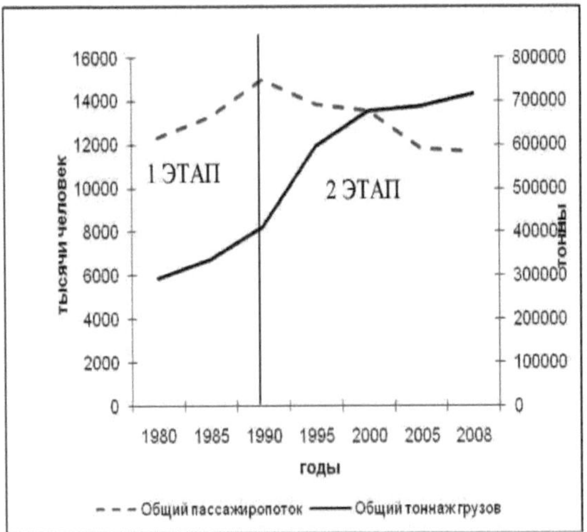


*Figure 9.* Trends in total freight and passenger traffic by major modes of transport in the Republic of Korea from 1980 to 2008 (Korea Statistical Yearbook, 2009). (from Korea Statistical Yearbook, 2009)

In order to qualitatively characterize the level of transport infrastructure in some territories of the Republic of Korea, a technical reliability coefficient (TSC) was calculated, which is the ratio of the sum of actual and theoretical speeds of various modes of transport [6, 48]. As a result, ATUs were identified where the actual travel time for passengers was longer than the theoretical time. These include Gangwon-do Province because of excessive tortuosity of highways, and all 7 metropolitan cities because of high traffic congestion, especially in the capital city. In Seoul, the average

The time spent by motorists on a single trip is 62 minutes, while in Daejeon it is 51 minutes (Figure 10). The opposite is the case in Gyeonggi-do, where due to excessive traffic density, motorists spend an average of 38 minutes per train [60].

*Figure 10.* Differentiation of the transport reliability coefficient of the ATU
of the Republic of Korea in 2009. (compiled by the author on the basis of

The graph of the main transport hubs, based on the principle of shortest distance, forms a pentagonal star on the territory of the country
without a northern peak (Fig. 11) [48]. On the basis of its analysis were calculated [6]:

- *the Goltz factor*, representing: $KG = \dfrac{l}{\sqrt{SN}}$ (11),

where $l$ is the length of the network, $S$ is the developed area, $N$ is the number of points, revealed a relatively high level of provision of the country's territory with the main types of transport routes (1.38).

- *the Engel coefficient*, calculated as: $KE = \dfrac{l}{SHH}$ (12),

where $l$ *is* the length of the network, $S$ *is* the developed area, $H$ *is* the number of vehicles, showed a very low average per capita transport availability of the population (0.0018).

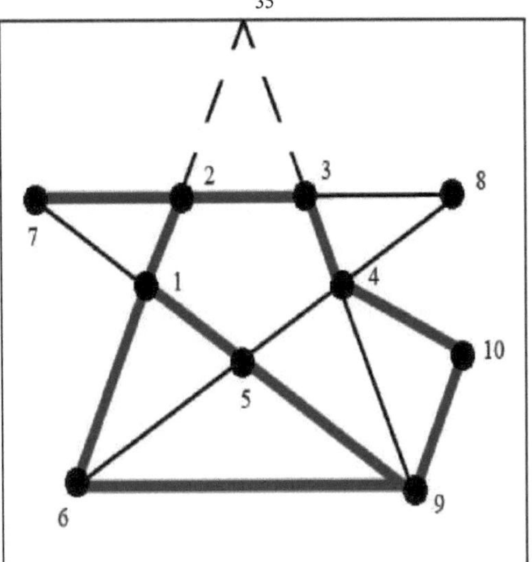



*Figure 11:* Diagram of the main transport hubs and highways of the Republic Korea, built according to the shortest distance principle (according to [60])

The typology of ATU of the Republic of Korea according to the level of transport system development was conducted on the basis of scoring the number of transport modes, number of tops, number of people employed in transport, as well as the number of transport enterprises (Table 9). These indicators correlate well with the main economic indicator of the territory - gross regional product per capita [27].

As a result of scoring the main indicators of transport system development in the Republic of Korea, three types of ATUs can be distinguished (Figure 12): highly, medium and low-developed transport system [60]. The type with a highly developed transport system includes the city of Seoul. Seoul and the Gyeonggi-do metropolitan area, together constituting the so-called *"metropolitan region"*. In spite of the lack and underdevelopment of air and water transport, the ATU has the highest number of transport companies (9,944 in Seoul and 6,2793 in Gyeonggi-do) and, consequently, the highest number of revenues from transport companies ($63,257,603 and $78,01854 respectively). In addition, more than 5% of the region's economically active population is employed in transport enterprises, i.e. exceeds the national figure by almost 2% (67).

Ta b l e 9

Scoring the level of development of the transport system of the Republic of Korea according to its key performance indicators (based on [67])

| ATE | Number of modes of transport (1) | Score 1 each | Transport company revenues (2) | Score 2 each | Number of transport enterprises (3) | Score by 3 | Number of people employed in transport (4) | Score by 4 | Amount of points | Type URTS |
|---|---|---|---|---|---|---|---|---|---|---|
| Seoul | 2 | 1 | 63257603 | 3 | 92944 | 3 | 368959 | 3 | 10 | 3 |
| Busan | 4 | 3 | 7136159 | 2 | 29101 | 1 | 103751 | 2 | 8 | 2 |
| Tag | 3 | 2 | 2553252 | 1 | 20247 | 1 | 58040 | 1 | 5 | 1 |
| Incheon | 4 | 3 | 4735749 | 2 | 19324 | 1 | 56504 | 1 | 7 | 2 |
| Gwangju | 3 | 2 | 1232370 | 1 | 9608 | 1 | 25638 | 1 | 5 | 1 |
| Taejon | 3 | 2 | 2952032 | 1 | 9692 | 1 | 30826 | 1 | 5 | 1 |
| Ulsan | 4 | 3 | 1180176 | 1 | 6414 | 1 | 19939 | 1 | 6 | 1 |
| Gyeonggi-do | 4 | 3 | 7801854 | 2 | 62793 | 2 | 153206 | 2 | 9 | 3 |
| Gangwondo | 4 | 3 | 905236 | 1 | 9814 | 1 | 23118 | 1 | 6 | 1 |
| North Chuncheon | 3 | 2 | 1404221 | 1 | 10399 | 1 | 25206 | 1 | 5 | 1 |
| Southern Chuncheon | 4 | 3 | 1329818 | 1 | 10882 | 1 | 27886 | 1 | 6 | 1 |
| North Cholla | 4 | 3 | 1272582 | 1 | 12014 | 1 | 29902 | 1 | 6 | 1 |
| South Cholla | 4 | 3 | 1542287 | 1 | 8914 | 1 | 31434 | 1 | 6 | 1 |
| North Gyeongsang | 4 | 3 | 2389812 | 1 | 16588 | 1 | 42272 | 1 | 6 | 1 |
| Southern Gyeongsang | 4 | 3 | 2245527 | 1 | 17225 | 1 | 41273 | 1 | 6 | 1 |
| Jeju-do | 4 | 3 | 491584 | 1 | 5829 | 1 | 12300 | 1 | 6 | 1 |

*Note.* The numbers in the "scores" and "type of FRM" columns indicate: 1 - poorly developed, 2 - medium developed, 3 - highly developed transport systems.

The second type includes two major economic centres, Busan and Incheon. Transport companies' revenues are more than 10 times lower than in the capital (USD 7136159 and 4735749 respectively).

The number of transport enterprises in these metropolitan cities is also noticeably inferior to the ATEs of the first type - 29101 and 19324 respectively [67]. The proportion of the economically active population employed in transport enterprises is lower relative to Seoul, but higher than the Republic of Korea average (3.6% and 3.3% respectively).

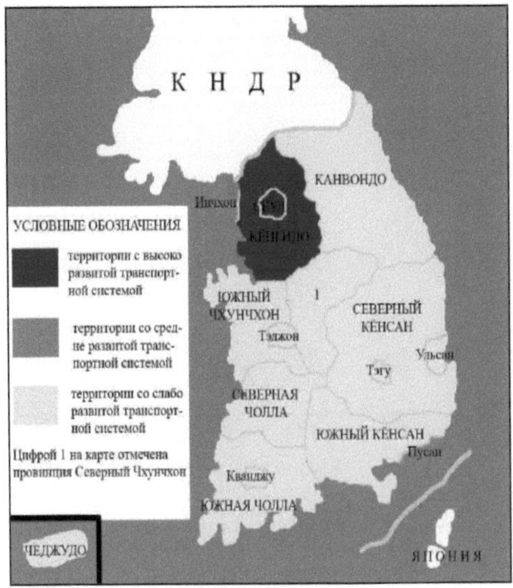



*Figure 12:* Typology of ATUs in the Republic of Korea by transport system development in 2009. (compiled by the author on the basis of calculated data)

Most of the ATEs belong to the latter type - territories with poorly developed transport system. They are inferior to the first two types in all considered indicators except the quantity of transport. Often prevailing in terms of the number of the main types of transport, they are inferior in terms of its qualitative development. All the above proves the relatively underdeveloped transport system of the Republic of Korea as a whole [60].

### 3.3. Agglomerations, their structure and evolution

***An urban agglomeration*** is a compact cluster of settlements, mostly urban, in some places coalescing, combined into a complex multi-component dynamic system with intensive production, transport and cultural relations [31].

Analysis of the population dynamics of major agglomerations in the Republic of Korea (Figure 13) showed that a significant increase in the intensity of large-city settlement growth has been observed since the urban transition, which took place between 1975 and 1980 (Em, 2009, b). (Em, 2009, b). However, the predominant growth of Seoul was marked immediately after the end of the Korean Civil War in 1953 (50).

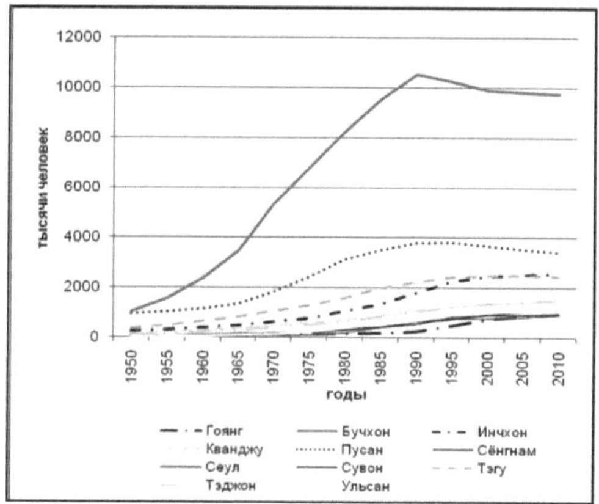


*Figure 13:* Population trends in the major agglomerations of the Republic of Korea from 1950 to 2010 (see [71]). (from [71])

### 3.3.1. Seoul as a world city and the country's largest agglomeration

About 82% of the population of the Republic of Korea today lives in cities. The rise in the proportion of the urban population was due to the industrialisation policies of the 1960s and the migration of people from the countryside to the cities. For a long time, Seoul has been the main centre of gravity for rural residents: 10% of the country's population lived there in 1960 and 23% in 2009 (69). - 23% [69]. And even despite the emergence of new centres of migration gravity (Busan, Pohang, Ulsan, Masan, Changwon), the population share of the Seoul agglomeration is still increasing. It now has a population of 23.9 million, making it the 2nd largest agglomeration in the world following Tokyo. The capital and the surrounding province of Gyeonggi-do are now home to about 49% of the country's population [66].

It is worth noting that Seoul, as a tier-2 global city, is a nodal element of the global settlement framework. In 2008, its population was 9,762,000, making it the 22nd largest city in the world.

**Seoul** has emerged as a powerful industrial metropolis that includes Incheon, South Korea's largest port on the Yellow Sea, as well as the cities of Goyang, Bucheon, and Suwon in the immediate vicinity of the capital [15]. The city has expanded dramatically in the process of urbanization and industrialization and continues to grow as a thriving and prosperous centre of Korea's political, economic, cultural, and educational life. Seoul agglomeration occupies a large part of Gyeonggi-do Province, which stands out in terms of the concentration of industry and the volume of not only industrial production but also in the provision of services. It is this circumstance that has contributed to making Seoul an important transportation hub.

Seoul's rapid growth began during the colonial era and continued during the years of the 'economic miracle'. From 1993 onwards, the population of the capital began to decline and continues to do so today. The population of the Korean capital decreased by 9% between 1993 and 2009, from 10.889 million to 9.895 million (67). It can be explained first of all by the intensive construction of "satellite cities" around the capital. The fact is that "administrative Seoul" is much smaller than the real Greater Seoul. The boundaries of the city have not been revised for a long time, and most of the new residential areas, not to mention the satellite cities, have formally fallen outside the capital [30]. Therefore Koreans themselves usually speak of a 'metropolitan area', which can be thought of as a circle about a hundred kilometres in diameter, centred on the southern border of Seoul proper. In 2009 it was home to 23.9 million people, or 46.8% of the total population (67). The Seoul metropolitan area is a dense concentration of of intellectual, economic, and political life in Korea (30).

Seoul is the centre of the metropolis, where almost half of all Koreans live, but the population distribution within its boundaries is not very even and requires detailed analysis.

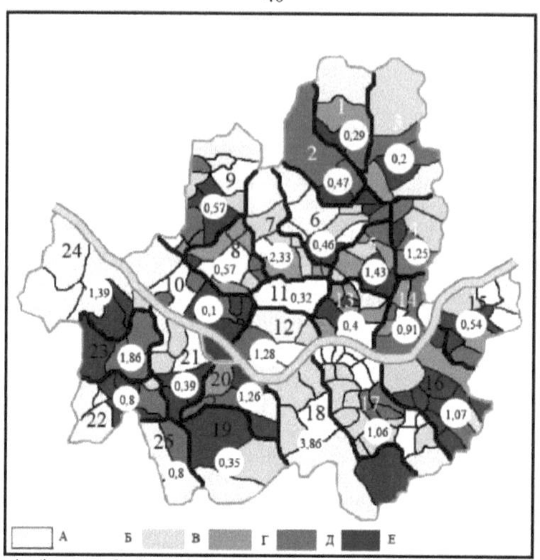



*Note.* Population density (in persons/km2): A - less than 10; B - from 10 to 15; C - from 15 to 20; D - from 20 to 25; E - from 25 to 30; F - over 30. In the circles by city districts the measure of territorial concentration of population is marked The numbers indicate the city districts: 1 - Dobong, 2 - Gangbuk, 3 - Novon, 4 - Junnan, 5 - Dongdaemun, 6 - Seongbuk, 7 - Jeongno, 8 - Seodaemun, 9 - Eupeng, 10 - Mapo, 11 - Jan, 12 - Yeongsan, 13 - Seondeong, 14 - Gwangjin, 15 - Gangdong, 16 - Songpa, 17 - Gangnam, 18 - Seocho, 19 - Gwanak, 20 - Dongjak, 21 - Youngdongpo, 22 - Guro, 23 - Yangcheon, 24 - Gunseo, 25 - Gemcheon.

*Figure 14:* Differentiation of population density and measures of spatial concentration in Seoul (compiled by the author using estimated data)

A modern map of Seoul (Figure 14) shows that the Hangang River, flowing from east to west, crosses the city roughly in the middle. If Seoul is represented as three concentric circles, the middle figure is the most densely populated. The calculations of the rank correlation values show that there is a fairly close relationship between the proportion that the districts occupy in area and in population in the city. The lowest average population density figure is found in Jongno district (6.93 persons/km2), which is dominated by parkland; the highest is in western Yangcheon (27.88 persons/km2) due to the dense urban development of high-rise residential buildings. There is considerable differentiation in the average population density within the districts themselves. The map most vividly describes the differentiation of the indicator by sub-districts of Seoul (Figure 14) [69].

Analysis of the spatial concentration of population in Seoul's districts, calculated using the formula:

$$K = \wedge|DSi - DPi| \ (13),$$

where $DPi$ is the share of population in the total population, $DSi$ is the share of i-territory in the total area, allows us to determine the orderliness of their settlement [46]. Jongno (2.33) and Seocho (3.86) districts are characterized by practically chaotic settlement, because they have a value of the analyzed index above 2. The districts that have a value of the indicator close to zero are characterized by an almost even settlement: Mapo (0.1), Nowon (0.2), and Seongdong (0.4) (69).

The centre of Seoul, which has a regular layout, is built up with high-rise, high-rise apartment blocks. The southern and eastern parts are dominated by narrow, crooked streets and the outskirts by slums. The main streets are Chonno and Sejonno. The city centre is dominated by the high-rise towers of the headquarters of major corporations, financial firms, banks and modern hotels. The business centre of the city is located to the north of Mount Namsan [15].

Seoul is the main academic and cultural centre of the Republic of Korea. It is home to some of the country's best universities, including Seoul University, Korea National University and Yeongseo University. The city is also home to major theatres and many interesting cultural sites.

### 3.3.2. Agglomerations and metropolitan cities

**Busan** is the country's second largest city and its largest port, home to more than 3.6 million people. It is located in the extreme southeast of the peninsula and is the maritime gateway to Korea. Busan was one of the first Korean ports to open to foreign ships in the late nineteenth century.

The city emerged in the first centuries AD as a settlement called Kaya (the same name as the state that existed there). In the 15th century Busan became an important seaside fortress and the centre of limited Japanese-Korean trade relations. After the colonization of Korea by Japan, Busan became the first port where Japanese merchants and businessmen settled. From 1925 to 1963 Busan was the administrative centre of the province of South Gyeongsang, after which it became a special purpose city, effectively becoming a separate administrative-territorial unit (50). Since the 1930s, it became a major industrial centre [15].

The modern city is a multi-storey development. It has developed mechanical engineering (shipbuilding, metallurgy, electrical industry) and textile industry.

Today, the population of the Busan agglomeration totals 3.97 million. It is ranked 102nd in the hierarchy of agglomerations of the world. The distribution of the economy and population gravitates predominantly towards the coast.

**Daegu** is the third most populous city in the country. It is located in the southeast and is one of the largest cities in the Republic of Korea in terms of area (885.56 square kilometres). The city, with a population of more than 2.5 million as of 1 January 2009, is divided into nine districts [67].

Since the 1960s, Daegu has been one of the country's largest export production centres. Daegu has become a link between the nearby specialised industrial centres of Gumi, Pohang, Ulsan and Changwon. The city has developed precision engineering, automobile manufacturing, semiconductor manufacturing, and more. In addition, 40% of the world production of polyester fibre and 43% of all domestic textile products are produced here [66].

Daegu has preserved a significant number of cultural and natural monuments located within the city limits. The city has six parks and four recreation areas. The city has six parks and four recreation areas and a national museum.
The city administration is implementing a 10-year plan to turn it into a centre for culture and the arts.

**Ulsan is** one of the largest industrial and transportation centres in the country. The area that Ulsan occupies has developed rapidly since 1962, when the county centre and its surrounding municipalities with a population of 80,000 were designated as a special industrial area as part of the country's modernisation programme. In 1997, Ulsan received the status of a special purpose city thanks to its progressive development.

As Ulsan was industrialised, administrative reforms expanded its area to 1,055.7 square kilometres, 1.7 times the area occupied by Seoul. Ulsan has a population of 1.11 million [67]. Ulsan's share in the industrial production of the Republic of Korea is 13%. In 2009, the city's exports abroad exceeded $18 billion and imports exceeded $19.5 billion [67]. The largest exports are chemicals and refined petroleum products, automotive, shipbuilding and iron and steel, as well as electronics, machinery and equipment. Imports include crude oil and raw materials for the chemical industry, machinery and equipment, ferrous metals and iron ore (15).

The city is home to two major national industrial complexes, Ulsan-Mipos and Ocean City. They house the enterprises of 39 corporations with investments from 10 countries with a total value of more than 800 million U.S. dollars [67]. The production facilities of the Hyundai Corporation are concentrated here, with its car-making, shipbuilding, ship-repair, pipe-rolling, paint-and-paint and other enterprises.

The port plays a crucial role in the life of the city. Ulsan's specialisation in handling bulk cargo, refined petroleum products and petrochemicals has made it a vital transport hub for the country. The most important sectors of the entire national economy are directly dependent on its smooth operation.

**Gwangju** is the administrative, industrial and cultural centre of South Jeolla Province and covers an area of 501.2 square kilometres. The city had a population of 1.42 million in 2009, ranking 5th in the hierarchy of cities in the Republic of Korea. Despite acquiring special purpose city status in 1986, it still serves as the administrative centre of South Jeolla Province due to the lack of an alternative centre in the province [25, 67].

The main focus of the city's economy is the development of medium and small businesses. The main industrial capacity is concentrated in the Hannam Industrial Complex, which came into operation in the early 1980s and is engaged in the processing of agricultural products. The expansion of the Pendon Industrial Complex and the High Technology Complex have increased the industrial capacity.

Gwangju is one of the world's centres of fibre optics, optical equipment and materials. The city has a development programme for communications equipment, lasers and the ceramics, plastics, solar cells and optical memory devices used in them [25].

The city is renowned for its cultural traditions, and a number of famous historical cultural monuments have been preserved there.

The population of **Daejeon** in 2009 was 1.49 million (67). Despite acquiring special-purpose city status in 1989, it is the administrative centre of South Chungcheong Province. Located practically in the geometric centre of the country, the city has an advantageous geographical position (Section 3.1.2.) because it lies at the intersection of the country's main polyhighway linking Seoul and Busan.

Daejeon is a city of science. It is home to many research institutes (Institute of Atomic Energy Research, Radiation Research Centre, Institute of Radiation Research for Agriculture). The city has a major centre for paper industry and silk weaving, cotton, food processing, and machine building [25].

### 3.3.3. Pendular migration as an indicator of labour integration

Pendular migration analysis can serve as an integral part of the analysis of urbanisation processes. It also helps to assess the degree of labour integration between labour donors and labour receptors.

Even before independence, the south of the Korean peninsula witnessed an increase in the intensity of daily labour trips, the bulk of which were rural and suburban residents eager to move to major cities (66). After the urban transition in 1977, the total volume of commuting increased as cities needed manpower (Figure 15) (63).

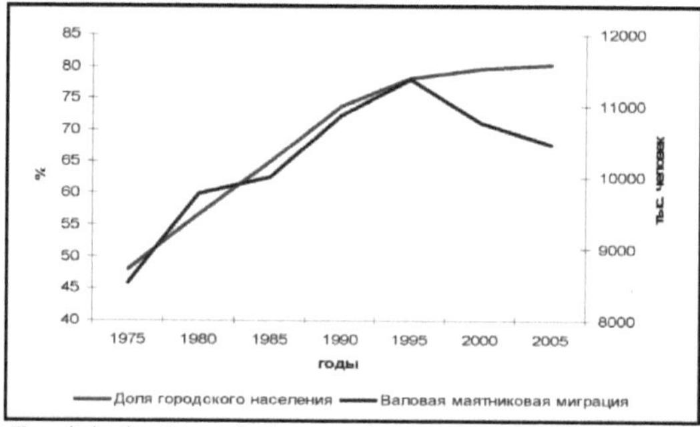


*Figure 15.* Trends in the share of urban population and gross commuting volume in the Republic of Korea from 1975 to 2005.

(from [66])

An analysis of the number of pendular migrants arriving and departing from the country's largest cities suggests that

- The number of incoming labour force in the large cities of the Republic of Korea, mostly composed of low-skilled people, has been gradually declining since 1995, thanks to government urban policies and the development of settlements adjacent to agglomerations.

- The number of daily commuters in major cities is decreasing in Seoul and Busan, and gradually increasing in Incheon, Daejeon, Daegu, Gwangju, and Ulsan. The bulk of the out-migration from these cities is made up of highly skilled people, the number of which is increasing every year due to their dynamic development.

The outflow of professionals from Seoul and Busan has slowed due to increased competition in the highly skilled labour market [63, 66].

*Pendular migration balance is* the difference between inbound and outbound pendular migrants carried out over a certain period of time within a specific territory [20, 47]. Its study in the Republic of Korea has grouped administrative-territorial units into the following groups [63]:

1    group - *administrative-territorial units (ATUs) with unstable dynamics of pendular migration balance* are characterised by a non-stable indicator (in Gyeonggi-do, Gangwon-do, South Chungcheong, North and South Jeolla, North and South Gyeongsang, Jeju-do, and in Seoul, Busan and Daejeon cities);

2    group - *the ATEs with steadily increasing pendular migration dynamics* include Daegu, Incheon, and Gwangju, which increase their attractiveness to labour force from surrounding provinces. Daegu and Gwangju are the only major cities in their surroundings;

3    The *ATU* group *with a steadily decreasing commuting balance* is represented by Ulsan City and North Chungcheong Province, which was the last to make the urban transition (1980 - 1985). An increasing number of its inhabitants are becoming pendular migrants. Most of them are moving towards the cities of Gyeonggi-do and South Chungcheong (Figure 16).

*Figure 16.* Sustainability of pendular migration balances in the territories of the Republic of

Korea (compiled by the author using estimated data)

A study of pendulum migration identified ATEs with positive and negative balances. Positive balances are characteristic of Seoul, Busan, Incheon, Daejeon and Gyeonggi-do. It is these ATEs that, from an economic point of view, are attractive to commuting migrants because of the significant demand for labour and higher wage levels compared to provincial cities. The ATEs belonging to the second group can be divided into two subgroups, taking into account the sustainability of the change in the balance of the indicator under study:

2A - to *ATEs with persistently negative swing migration balance* belong to the category of permanent "donors" of labour for type 1 provinces and cities (Daegu, Gwangju, Ulsan, North and South Jeolla, North and South Gyeongsang, Jeju-do).

2B - *ATEs with unsustainable negative balances* are Gangwondo, North and South Jeolla [63].

The time it takes pendular migrants to travel from their places of residence to work [55] in the largest cities of the Republic of Korea is always longer than that spent by residents within the city itself (Figure 17).

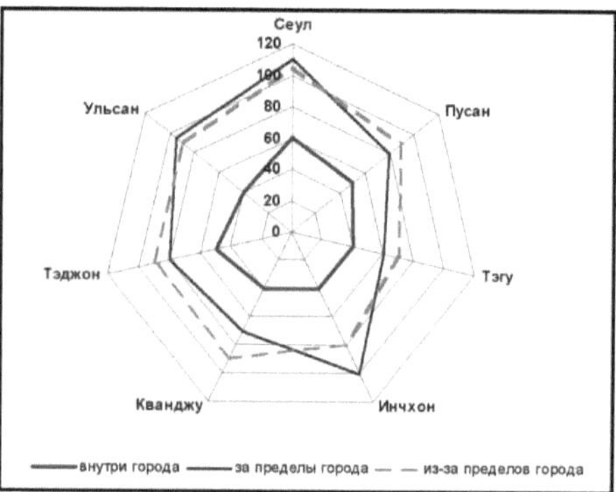


*Figure 17:* Systemic Time Use for Labour Travel (compiled by the author using calculated data)

To analyze the nature of the connection between the central city and its zone of influence within the largest agglomerations, we calculated the indicator "M": where $Pr,Pc$ - the number of employed people working in the central city or suburban zone; $Nr,Nc$ - the number of employed people living in the central city and suburban zone [17, 18]. It allowed revealing the degree of labour integration of central cities and their suburban zones. On the average it was 26.7% in 7 largest agglomerations of the country in 2009 (Table 10), but the range of its variation is significant: from 17.6% in Daegu to 36.4% in Daejeon [63]. An attempt was made to classify agglomerations according to the degree of integration. For this purpose, the following groups were identified: those with a high degree of labour integration of suburban zones and central cities (Incheon, Daejeon), with a medium degree (Seoul, Busan, Ulsan), and with a low degree (Daegu, Gwangju).

$$\frac{Prc + Pcr}{Nr + Nc} \quad (14),$$

Table 10

Degree of labour integration of the largest agglomerations and their peri-urban areas in the Republic of

Korea in 2009. (compiled by the author using estimated data)

| № | Agglomeration | Degree of labour integration of agglomeration and peri-urban area, % | Agglomeration Group by degrees of integration |
|---|---|---|---|
| 1 | Seoul | 31,6 | 2 |
| 2 | Busan | 25,8 | 2 |
| 3 | Incheon | 31,8 | 1 |
| 4 | Agglomeration Tag | 17,6 | 3 |
| 5 | Taejon | 36,4 | 1 |
| 6 | Ulsan | 23,4 | 2 |
| 7 | Agglomeration Gwangju | 20,4 | 3 |

*Note.* The numbers denote the following groups: 1 - increased, 2 - medium, 3 - decreased degree of labour integration of agglomeration and suburban areas.

The analysis of the private spatial indicator of pendulum mobility in metropolitan cities was calculated as:

$$P = \frac{W}{N} * 1000 \qquad (15),$$

where $W$ is the total value of all pendulum migrants moving in the system, $N$ is the total number of workers and employees living in the settlement system [47]. It showed that Daejeon and Incheon have the highest mobility, with figures exceeding 140‰. The lowest value is characteristic of Busan, less than 74‰ [63].

Pendular migrants constitute a varying share of the labour force in their places of work (Figure 18). In 2008, their share was high in the metropolitan cities of the capital region - Seoul (14%), and Incheon (12%), due to its 24 urban settlements; and Daejeon (12%), the only major city at the junction of 3 provinces. Busan (11%) and Gwangju (10%) are average.

Ulsan (8%) and Daegu (7%) have the lowest share.

*Figure 18:* Proportion of circular migrants in the total labour force

(compiled by the author on the basis of calculated data)

An estimate of the balance between jobs and the number of people living in the labour force has been calculated using the formula:

$$Mt = \qquad\qquad (16),$$

where $Ra$ is the number of workers and employees living and working in settlement $t$; $Rs$ is the number of workers and employees arriving daily at $t$ *to* work from other settlements; $Rt$ is the number of workers and employees leaving daily from $t$ *to* work in other settlements (Taborisskaya, 1979). It allowed for the identification of territories with vacancies (Seoul,
Busan, Incheon, Daejeon), and a shortage of places of employment (Daegu, Gwangju, Ulsan). The analysis proves that highly urbanized areas in the north-west and south-east of the country are highly attractive for labour migrants (Figure 19) [63].

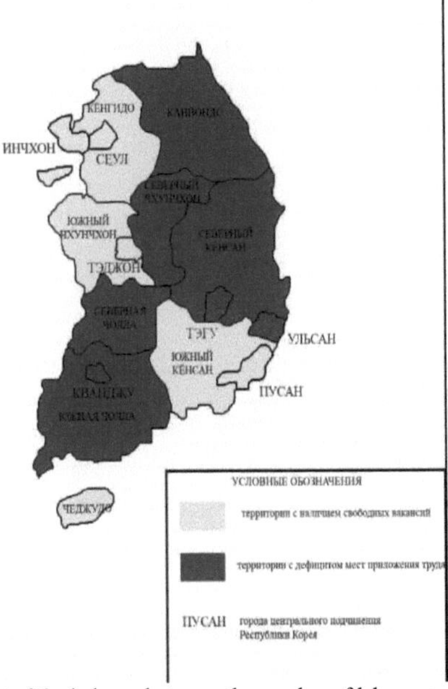


*Figure 19:* Assessment of the balance between the number of labour resources and places

of work (compiled by the author on the basis of calculated data)

## Chapter 4: Evolutionary processes in the Republic of Korea's urban settlement system

### 4.1 Changes in the hierarchical structure of urban settlement

The current network of cities in the Republic of Korea comprises 84 settlements. Its important feature is the relatively even development of the whole system, as well as its increasing internal orderliness.

The best basis for analysis is the movements of the units that make up the mix of cities by tens (Table 11) [54]. The degree of regularity of size and rank distribution, as well as shifts in this distribution over time, served as the basis for identifying trends in the development of the urban system in the late twentieth century. - beginning of the XXI century. The analysis showed that the period from 1990 to 1995 was more shifted than the period 1995-2000, and that the shift between 2000 and 2008 was more regular than in the previous five-year period [61].

Table 1 1

Changes in the ranking by size of cities in the Republic of Korea from 1990 to 2008. (based on [53, 61]).

| Rank cities in 1990. | Number of cities with changing ranks (by number of ordinal numbers) | | | | |
|---|---|---|---|---|---|
| | 0-5 | 5-10 | 10-15 | 15-20 | Over 20 |
| 1-10 | 9 | 1 | 0 | 0 | 0 |
| 11-20 | 5 | 4 | 1 | 0 | 0 |
| 21-30 | 4 | 3 | 2 | 1 | 0 |
| 31-40 | 2 | 2 | 5 | 0 | 1 |
| 41-50 | 3 | 2 | 2 | 1 | 2 |
| 51-60 | 1 | 0 | 4 | 3 | 2 |
| 61-70 | 2 | 3 | 4 | 0 | 1 |
| 71-80 | 3 | 3 | 1 | 1 | 2 |
| 81-84 | 3 | 0 | 0 | 0 | 1 |

When considering changes in the ranking of cities over the period under study, large movements of individual components with different directions and intensities are noticeable in almost all dozens of cities. Their movement reflects the dynamic equilibrium of the system with the large strength of some impulses. The result is certain shifts in the regularity of individual distributions. The exception was the first dozen cities, whose ranking was very stable due to the presence of millionaire cities, with some deviations at the end of the series.

Minimal changes in ordinal numbers are observed in the cities of the first, second and eighth tens.

There is a high level of spatial concentration in the millionaire cities and a high degree of hypertrophy in the capital: Seoul has a share of more than 21 % [61].

The gap between the values of the first and second largest cities is quite large. In 2008, the populations of Seoul and Busan correlated to each other as 1:2.9, which proves the primate type of distribution (Figure 21, c). In the distribution between the second and subsequent components, the relationship graph shows a relatively even distribution of the size of the cities in the system.

As a result of the upward movement of many small and medium-sized cities with high population growth rates, the number and proportion of units in these classes have decreased. There has been no increase in the size of the smallest city due to the acquisition of urban status by Geryong Township in South Chungcheong Province between 2000 and 2005. The difference between the smallest and largest city in the system has decreased from 10.9 to 10.4 million people. This proves an emerging trend of stabilization of this indicator (61).

Particularly dramatic changes in ranks were found mainly in cities in the fifth and eighth tens (Gwangju - 36 positions up, Yeongjin - 26). The cities with extreme decreases in population ranks are predominantly located in the centre of the country (Junggyeop and Sangju - 25 and 19 points, respectively).

### 4.2 Sustainability of the distribution of urban settlements

Settlement systems tend to maintain a rank-size distribution (formula 1) in a certain way as the proportion of the urban population changes. The work of A.A. Vazhenin proved that the hierarchical structure of most countries and regions of the world best fits the Zipf rule Zipf rule at a 50% level of urbanization. However, this provision was not confirmed for the study country [10].

Figure 20 shows the trends of the real and ideal hierarchical distribution of the ten largest cities in the Republic of Korea, according to the Zipf rule in 1950. (a), in 1980. (b), and in 2010. (в). In this case, we observe the pattern that: the higher the proportion of urban population, the more compliance with the rank-size rule increases.

There is no principle hypertrophy of Seoul in 1950. The second largest seaside city, Busan, has a population exceeding the "ideal" value by almost 100,000. However, cities ranked 3 and below lag noticeably behind the values according to the Zipf rule.

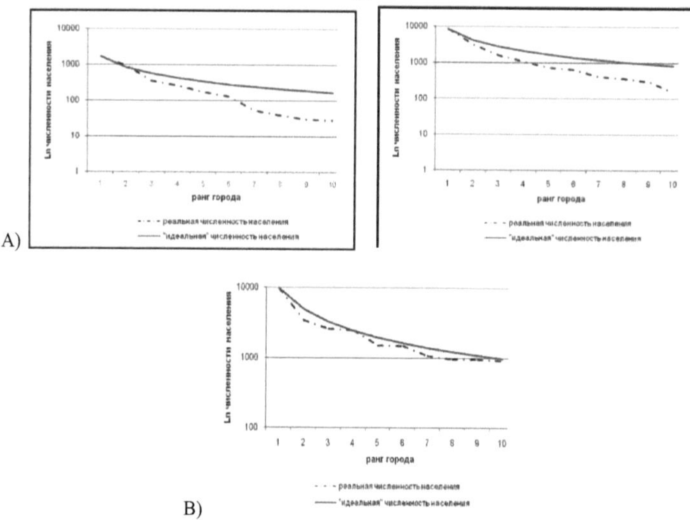

A)

B)

*Figure 20:* Change in the location of the ten largest cities in the Republic of Korea according to the rank-size rule in 1950. (A), 1980. (B) and 2010. (C) (according to [71])

In 1980, after the urban transition, there is an increase in the hypertrophy of Seoul. It becomes 2.7 times larger than the second largest city. In addition, Busan lags behind the ideal city by almost 1 million inhabitants.

The "rank-size" rule. In doing so, there is a gradual approximation of the trend of the real city distribution to the "ideal" one at the end of the series.

Finally, in 2010, the Republic of Korea's settlement system is characterized by a relatively high hierarchical ordering of the top 10 cities. Of the 9 cities following Seoul, 3 are larger or almost equal to the 'ideal' size.

Comparing Fig. 20-a and Fig. 20-b we can get a confirmation of the judgement that "the higher the level of urbanisation, the greater the number of the largest cities in the system and the greater the deviation from the ideal" [10]. However, the trend of 2010. (Fig. 20, c) does not obey this judgement. This is due to the launch of urban policies that have relatively reduced the excessive hypertrophy of Seoul, thereby unloading it.

This change in rank-size trends in the system studied is complemented by another indicator, namely:

$$\frac{\ln Pt}{\ln V} \quad (17).$$

In all cases a value close to 0.9 was obtained, which A.A. Vazhenin designated as *Co*. Analysis of the *Coe* indicator (Table 12) showed that from 1950 to 2010, despite a significant increase in both the number and the share of urban population, its values are very close to the "ideal" and the amplitude of differentiation from the value of 0.9 does not exceed 0.04, that is, *Coe* has not undergone serious changes for half a century [9].

This proves that the *Co* indicator is absolutely adequate and indicates that there is a systemic link between urban settlements in the country.

Thus, the rank-size distribution of cities in the Republic of Korea, despite a significant increase in urbanization, has shown an invariant proximity to a certain indicator for a long time, which means that we are able to compare them with systems of central places. This also proves that there is an invariant that sets the size of urban settlements in settlement systems and is determined only by the size of urban population [9].

Table 1 2

Dynamics of real and hypothetical population of the main city in the settlement system of the Republic of Korea from 1950 to 2010  (compiled by the author using estimated data).

| The year | Proportion of urban population, % | Mortgage value of the main city, thousand people | Population of the largest city, thousand people | From |
|---|---|---|---|---|
| 1950 | 21,4 | 2492 | 1693 | 0,94 |
| 1955 | 24,4 | 2745 | 1574 | 0,92 |
| 1960 | 27,7 | 3110 | 2445 | 0,9 |
| 1965 | 32,4 | 3859 | 3471 | 0,9 |
| 1970 | 40,7 | 5436 | 5433 | 0,9 |
| 1975 | 48 | 7110 | 6889 | 0,91 |
| 1980 | 56,7 | 8976 | 8364 | 0,91 |
| 1985 | 64,9 | 10146 | 9639 | 0,9 |
| 1990 | 73,8 | 11051 | 10613 | 0,89 |
| 1995 | 78,2 | 11109 | 10596 | 0,89 |
| 2000 | 79,6 | 11057 | 10373 | 0,88 |
| 2005 | 80,8 | 11029 | 10297 | 0,88 |
| 2010 | 81,9 | 11091 | 10464 | 0,87 |

## 4.3 Techniques for constructing a hexagonal lattice

Determining central locations by hierarchical levels is a complex task that requires a comprehensive approach [52]. However, in this case, the results of research by A.A. Vazhenin, who proved on the basis of numerous studies that ratios in the average size of cities at different levels of the hierarchy for any modification of the settlement system are 1.0 - 0.33 - 0.10 - 0.03 - 0.01 [10] were used. In accordance with the relativistic theory of W.A. Schuper, the Seoul agglomeration has been taken as the main central place, the size of which according to the latest data is about 23.9 million people [71]. Consequently, the population of cities at the second level of the hierarchy should be about 8 million people, but there are no such agglomerations in the country. This means that the second level of the hierarchy drops out in the settlement system. The reasons for this will be outlined in Section 4.4. However, at the third level of the hierarchy there are 4 agglomerations formed around Busan, Daejeon Daegu and Gwangju, with an average population size of about 10% of Seoul. At the same time, the size of the fourth-tier cities corresponds to 8 localities, most of which are located in the Capital Region (Table 13). The location of 4 cities in the third hierarchical level, and 8 in the fourth, despite the absence of the second, proves that the settlement system corresponds to $K = 3$. The rest of the evidence will be given in Section 4.4.

Table 13 Example of the distribution of settlements into hierarchical levels based on size

(compiled by the author on the basis of calculated data)

| № | Settled item | Population, thousand people | Average population (Ps), thousand people | P 1 / Ps | Level of hierarchy |
|---|---|---|---|---|---|
| 1 | Seoul agglomeration | 23900 | 1 | 1 | 1 |
| 2 | Busan agglomeration | 3615 | | | 3 |
| 3 | Daegu agglomeration | 2513 | | | 3 |
| 4 | Taejon agglomeration | 1488 | 2259 | 10,1 | 3 |
| 5 | Gwangju agglomeration | 1423 | | | 3 |
| 6 | Ulsan agglomeration | 1113 | 701 | 34 | 4 |
| 7 | Youngjin | 814 | | | 4 |

| 8 | Ansan | 735 | | 4 |
|---|---|---|---|---|
| 9 | Cheongju | 638 | | 4 |
| 10 | Anjang | 631 | | 4 |
| 11 | Cheongju | 627 | | 4 |
| 12 | Cheonan | 541 | | 4 |
| 13 | Changwon | 510 | | 4 |

To construct the hexagonal lattice, V.A.'s methodology was used. Schuper's paper "The Self-Organisation of Urban Settlement" [58].

For this purpose, a circle was drawn on which all 84 urban settlements of the Republic of Korea were plotted. An auxiliary line perpendicular to the meridians was drawn on a map of the country from the centre of Seoul to the east coast. Angles were measured between it and the rays crossing the points. Under this angle, increased by half, a radius was drawn on the cartoid shown in Fig. 21. On this ray the town, the central place, was plotted so as to divide it into two segments in the same proportion as this central place divided the ray drawn through it on the map (tab. 1, appendix). The island of Jeju-do was regarded as part of the landmass.

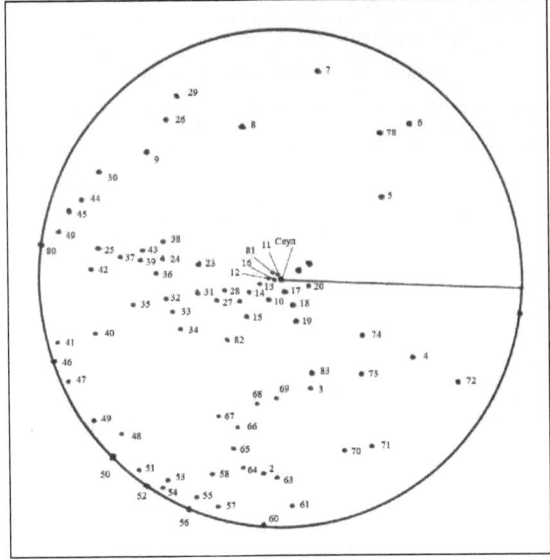


Cities that are part of the Seoul Metropolitan Area are not marked: 1-Seoul, 3-Goyang, 5-Bucheon, 6-Incheon, 12-Suwon, 17-Seongnam.

The numbers on the mapoid indicate the cities: 2 - Phaju, 4 - Gunpo, 7 - Siheung, 8 - Ansan, 9 - Hwasong, 10 - Euwang, 11 - Gwacheon, 13 - Osan, 14 - Pyeongtaek, 15 - Ansong, 16 - Yeongin, , 18 - Gwangju, 19 - Icheon, 20 - Hanam, 21 - Guri, 22 - Namyangju, 23 - Eujongbu, 24 - Yanju, 25 - Dongduchen, 26 - Pochon 27 - Cheonan, 28 - Asan, 29 - Seosan, 30 - Boreang, 31 - Konju, 32 - Nonsan, 33 - Gereng, 34 - Daejeon, 35 - Cheongju, 36 - Iksan, 37 - Kimche, 38 - Gunsan, 39 - Chongyub, 40 - Namwon, 41 - Suncheon, , 42 - Gwangju, 43 - Naju, 44 - Mokpo, 45 - Cheongnam, 46 - Yesu 47 - Gwangyang, 48 - Jinju, 49 - Sachon, 50 - Tongyeong, 51 - Koje, 52 - Masan, 53 - Changwon, 54 - Jinhae, 55 - Gimae, 56 - Busan 57 - Yangsan, 58 - Miryang, 59 - Gwangmeng, 60 - Ulsan, 61 - Gyeongju, 62 - Pohang, 63 - Yeongcheon, 64 - Gyeongsan, 65 - Daegu 66 - Gumi, 67 - Gimcheon,68 - Sangju, 69 - Mungyeong, 70 - Andong, 71 - Yeongju, 72 - Taebaek, 73 - Jechon,74 - Wonju, 75 - Samchok 76 - Donghae, 77 - Gangnam, 78 -

Chuncheon, 79 - Jeju, 80 - Seogwipho, 81 - Anyang, 82 - Cheongju, 83 - Chungju, 84 - Sokcho.

*Figure 21:* Cartoid depicting the transformed system of central places in the Republic of Korea (compiled by the author using calculated data from Table 1 in the appendix).

After that with the same radius according to the existing proportions a stencil of a hexagonal grid was prepared. In order to distribute cities on it, the stencil and the resulting circle with the applied points must be aligned and rotated until the points coincide as much as possible. After that, the cities are transferred to the layout of the country's hexagonal grid.

### 4.4 Analysis of the evolution of urban settlement systems

In the pre-war period, the Korean peninsula's unified settlement system was essentially extensive, in which a regular network of settlements was formed and stable communications between them were assured. Rural settlement corresponded to this type (50).

The end of the Korean Civil War in 1953 resulted in the formation of two sovereign states, whose settlement systems have since become fragmented. Compounding the near-total destruction of all settlements, including urban settlements that had been almost completely destroyed during the war, was the complete breakdown of all ties outside the demilitarized zone, which straddles the 38th parallel. The country, located in the south of the Korean peninsula, has become virtually isolated from the continent because of its location in the north of the Democratic People's Republic of Korea.

Immediately after the end of the war and throughout the second half of the twentieth century, the urban population in the south of the Korean peninsula began to grow rapidly. In a relatively small area, due to the destruction of Seoul and the rapid development of Busan, which was a link to international trade, a polycentric settlement system of two virtually independent systems of central places headed by Seoul (about 1.5 million people) and Busan (about 1.2 million people) with $K = 2$ was formed in 1955 (71). Despite the mismatch of city sizes with theoretical values, the sum of ratios of theoretical to empirical radii, has a value of 1.93, demonstrating almost perfect isostatic equilibrium (Table 14). As for the average distances between the cities at different levels of the hierarchy, the distance between levels I and II is twice as large as that between levels II and III (Table 14).

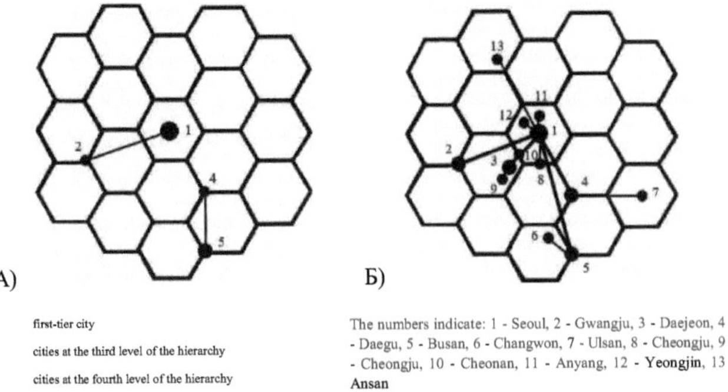



*Figure 22:* Evolution of the hexagonal lattice of the Republic of Korea (A - 1953, B - 2009) (compiled by the author using calculated data)

Table 14 Dynamics

of the main characteristics of the Republic of Korea's central location systems

(compiled by the author using calculated data)

| The year | Proportion of urban population | Type of system | Number of central seats on the level | Average population levels, thousand people | | | |
|---|---|---|---|---|---|---|---|
| | | | | 1 | 2 | 3 | 4 |
| 1953 | 21,4 | K = 2 | 1-1-2 | 1575 | 1046 | 201 | - |
| 1985 | 64,9 | K = 3 | 1-0-4-7 | 9639 | 0 | 895 | 312 |
| 2009 | 81,9 | K = 3 | 1-0-4-8 | 23900 | 0 | 2259 | 701 |

| | Distance, km | | | Radii | | | |
|---|---|---|---|---|---|---|---|
| | I-II | II-III | III-IV | $\dfrac{R1t2}{R1e2}$ | $\dfrac{R_{23}}{R\,2z}$ | $\dfrac{R.}{R_e}$ | $2 \dfrac{}{Re\,Re}$ |
| 1953 | 331 | 165 | - | 0,8/ 1,1 | 1,09/ 0,9 | - | 1,93 |
| 1985 | 243,3 | | 37,6 | 0,81/0,75 | | 0,63/0,82 | 1,85 |
| 2009 | 243,3 | | 26,5 | 0,81/0,75 | | 0,63/0,78 | 1,86 |

Population growth in the Republic of Korea's settlement system in

In the second half of the twentieth century, with a limited territory, the capital city began to significantly outpace all other cities in terms of growth. And the processes of agglomeration, and later megapolisation, which intensified in the 1980s, finally established Seoul as the centre without any alternative. This entailed the transformation of the settlement system into a monocentric modification with $K = 3$ in 1985. Because of the excessive population concentration in the Capital Region, which has about 49% of the population, the second level of the hierarchy has dropped out, because no urban agglomerations of the theoretical size could be found in the settlement system. The Seoul agglomeration has taken over their function. In this case, the ratio of the average size of cities at different levels of the hierarchy was 1 - 0 - 10.8 - 30.9, which almost perfectly corresponds to the proportions established by A.A. Vazhenin. The indicator determining the isostatic equilibrium is the maximum permissible value of 1.85. In 2009, due to the emergence of an additional city on the 4th hierarchical level, located in Gyeonggi-do province, the value of the average distance, and hence the empirical radius between the 3rd and 4th level cities decreases. Due to this, compliance with the isostatic equilibrium condition also increases to 1.86 (Table 14). In addition, the correspondence between the sizes of settlements of different levels has also increased and is as follows: 1 - 0 - 10.1 - 34 (Table 13).

### 4.5 Development forecast

The prognosis for the evolution of the urban settlement system of the country of study has two options. The first is its further independent development as a separate state, and the second is its development as part of a unified Korea. The development forecast was made for the year 2030. An exponential population projection formula was used to determine the size of cities and agglomerations:

$$P(t) = P_o * e^{rt} \quad (18),$$

where $Po$ is the initial population size, $r$ is the population growth rate, $t$ is the calculation time [1]. The results are presented in Table 3 in the appendix.

Let us consider the first option. According to the calculated data (Table 15) in the urban settlement system of the Republic of Korea as a sovereign country in 2030, due to the presence of agglomerations, the second level of the hierarchy will drop out as it does today. At the same time, it will retain the modification with $K = 3$, but a new city, Ulsan, has appeared on the third level due to its development. This can be proved by the value of the ratio of Seoul agglomeration to the average size of Tier 3 cities (Table 15), which is 11.1. Without Ulsan agglomeration it is 8.5 and with the next city in the hierarchy it is 12.1. Thus, the value of 11.1 corresponds to the established proportions as much as possible. Also, the fourth level of the hierarchy is marked by an increase in the number, in which 11 cities stand out, whose ratio of the average value to the Seoul agglomeration is 33.6. In addition, the isostatic equilibrium index, is equal to 1.89, which is also optimally close to the ideal value of 2, compared to the system modification with $K = 2$ and $K = 4$ (Table 17). This proves that there are five agglomerations on the third level of the hierarchy.

Forecast of the distribution of human settlements in the Republic of Korea in 2030 by level of hierarchy based on size (compiled by the author on the basis of calculated data)

Ta b l e 15

| № | Settled item | Population, thousand people | Average population (Ps), thousand people | P 1 / Ps | Level of hierarchy |
|---|---|---|---|---|---|
| 1 | Seoul | 24724 | 24724 | 1 | 1 |
| 2 | Busan | 2989 |  |  | 3 |
| 3 | Agglomeration | 2611 |  |  | 3 |
| 4 | Taejon | 2171 | 2209 | 11,1 | 3 |
| 5 | Agglomeration | 1773 |  |  | 3 |
| 6 | Ulsan | 1499 |  |  | 3 |
| 7 | Youngjin | 322 |  |  | 4 |
| 8 | Ansan | 1086 |  |  | 4 |
| 9 | Cheongju | 1073 |  |  | 4 |
| 10 | Anjang | 764 |  |  | 4 |
| 11 | Cheongju | 756 |  |  | 4 |
| 12 | Cheonan | 837 | 735 | 33,6 | 4 |
| 13 | Changwon | 516 |  |  | 4 |
| 14 | Pohang | 500 |  |  | 4 |
| 15 | Kyme | 790 |  |  | 4 |
| 16 | Namyangju | 812 |  |  | 4 |
| 17 | Ejonbu | 625 |  |  | 4 |
| 18 | Masan | 393 | 706 | 34,9 | 5 |

The hexagonal lattice undergoes a slight modification (Figure 23, a). However, most of the agglomerations that make up the third level of the hierarchy (Gwangju, Busan, Daegu, and Ulsan) are located in the edges of hexagons.

Tabl. 16

Projection of the distribution of settlements in the unified Korea in 2030 by hierarchy level  based on size (compiled by the author using estimated data)

| № | Settled Item | Population, thousand people | Average population (Ps), thousand people | P 1 / Ps | Level of hierarchy |
|---|---|---|---|---|---|
| 1 | Seoul | 24724 | 24724 | 1 | 1 |
| 2 | Pyongyang | 3630 | | | 3 |
| 3 | Busan | 2989 | | | 3 |
| 4 | Agglomeration | 2611 | | | 3 |
| 5 | Taejon | 2171 | 2445 | 10,1 | 3 |
| 6 | Agglomeration | 1773 | | | 3 |
| 7 | Ulsan | 1499 | | | 3 |
| 8 | Nampo | 1278 | | | 4 |
| 9 | Ansan | 1086 | | | 4 |
| 10 | Cheongju | 1073 | | | 4 |
| 11 | Hamhoon | 882 | | | 4 |
| 12 | Cheonan | 837 | | | 4 |
| 13 | Namyangju | 812 | | | 4 |
| 14 | Kyme | 790 | | | 4 |
| 15 | Anjang | 764 | 737 | 33,5 | 4 |
| 16 | Cheongju | 756 | | | 4 |
| 17 | Jeju | 643 | | | 4 |
| 18 | Ejonbu | 625 | | | 4 |
| 19 | Siheung | 519 | | | 4 |
| 20 | Changwon | 516 | | | 4 |
| 21 | Gumi | 511 | | | 4 |
| 22 | Pohang | 500 | | | 4 |
| 23 | Pyeongtaeg | 496 | 721 | 34,3 | 5 |

The second option for the development of the settlement system involves considering the entire Korean Peninsula. Here it is important to note that the projected size of the agglomerations located in the DPRK today was calculated by UN experts [71]. The population size of the remaining cities was calculated according to formula 18 (Table 2 in the Appendix).

The urbanisation rate will decrease slightly when the North and South are united. By 2030, it will be 81.4%. This is due to the lower degree of urbanization in the Democratic People's Republic of Korea.

Analysis of the size of agglomerations and cities on the Korean Peninsula in 2030 suggests that a unified settlement system will also correspond *to K* = 3. When the second level of the hierarchy drops out, 6 agglomerations are noted at the third level, with Pyongyang coming in second place, right after the metropolitan area. Table 13 shows that the ratio of Seoul agglomeration to the average value of third-level agglomerations, will be 10.1, which is only 0.1 more than the ideal value established by A.A. Vazhenin [10]. In this case, the fourth level of the hierarchy highlights 15 settlements, including the 2 largest agglomerations located north of the demilitarized zone. The ratio of the main central location to the average size of these 15 urban settlements is 33.5. The deviation from the ideal value is only 0.5 (Table 17).

The hexagonal lattice of a unified Korea in 2030. (Figure 23(b)), there is a marked increase in the complexity of the pattern. Four of the six third-tier agglomerations (Gwangju, Busan, Daegu, and Ulsan) are located at the tops of hexagons. Still a few cities, being the central locations of the fourth tier, are directly subordinated to the Seoul agglomeration.

Comparing the main characteristics of the settlement systems of the two proposed options, it should be noted that the average distance between the first and third tier cities in the forecast modifications decreased by more than 20 km, while, on the contrary, between the third and fourth tiers it increased (Table 17). This is explained by the relatively close proximity of the Pyongyang agglomeration to the Seoul agglomeration.

Due to the inclusion of the Pyongyang agglomeration into the country's population centres in the unified Korea option, the average number of third-tier cities is more than 400,000 more than that for  Republic of Korea (Table 17). This is due to the size of the Pyongyang agglomeration, which is expected to rank 2nd in the overall hierarchy of human settlements (Table 16).

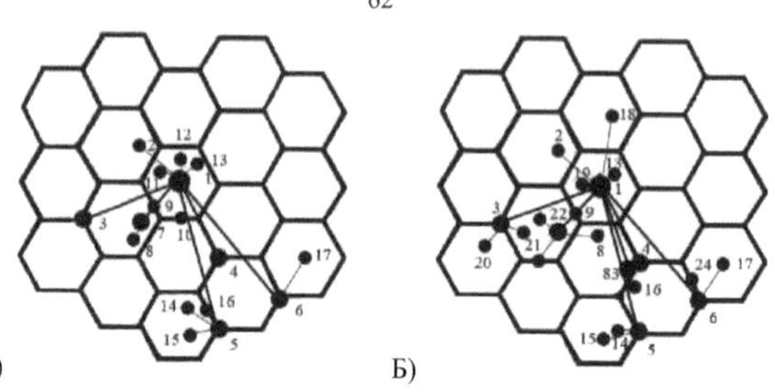



*Figure 23:* Forecast of hexagonal lattice development in 2030 for the Republic of Korea (a) and united Korea (b) (compiled by the author using calculated data)

Notably, the sum of the ratios of the empirical and theoretical radii of the different levels of the unified Korea hierarchy is closer to the ideal value than the Republic of Korea alone (Table 17).

Tabl. 17

Main characteristics of the projected settlement systems in 2030
(compiled by the author on the basis of calculations)

| The year | Proportion of urban population | Type of system | Number of central seats on the level | Average population levels, thousand people | | | |
|---|---|---|---|---|---|---|---|
| | | | | 1 | 2 | 3 | 4 |
| Korea | 81,4 | K = 3 | 1-0-6-15 | 24724 | 0 | 2445 | 737 |
| Republic of Korea | 86,3 | K = 3 | 1-0-5-11 | 24724 | 0 | 2209 | 735 |

| | Distance, km | | | Radii | | | $\frac{2}{Re}\overline{Re}$ |
|---|---|---|---|---|---|---|---|
| | I-II | II-III | III-IV | $Rn$ | $\dfrac{R_{2H}}{R_{2z}}$ | $\dfrac{R_{34}}{R_e}$ | |
| Korea | 212,6 | | 85,6 | 0,73/0,7 | | 0,79/0,83 | 1,99 |
| Republic of Korea | 216 | | 42,7 | 1,25/1 | | 0,25/0,39 | 1,89 |

**Conclusion**

The study showed that the Republic of Korea's urbanization is characterized by a "catching-up" pattern: low initial urbanization and a very rapid growth rate. Although the urbanization threshold was crossed in 1977, spatial differentiation of key process indicators still persists. The rapid growth of large cities has made agglomerations the main form of settlement in the country. The largest of these, the Seoul agglomeration, is a major element of the global settlement framework, firmly establishing itself as a world city.

An assessment of the EGP of the largest cities in the country has shown that almost all are characterised by high and medium EGP potential. The area as a whole has a relatively high potential for development. The maximum EGP potential is observed for cities gravitating towards the geometric centre. This means that this is where the new capital should be located in case of relocation.

The core of the ROCs are cities and urban agglomerations with more than 100,000 inhabitants. Most of them are concentrated in two highly urbanised areas: the Capital City and the South-East. The second most important linking element of RCDs are transport arteries. An analysis of the development of the country's transport system has shown a relatively insufficient level of development.

From 1985 to 1995, the process of suburbanisation became active in the country, accompanied by an increase in the capacity of pendular migration flows. Its decrease after 1995 proves that this process tends to subside. At the same time, a close labour link between agglomerations and suburbs has been detected.

The rank-size distribution of cities in the Republic of Korea over the past 60 years has shown a consistent proximity to a particular indicator, proving its resilience. At the same time, the metropolitan agglomeration remains hypertrophied. For this reason, the second tier of cities drops out in the country's settlement system. The indicators introduced in the relativistic central place theory prove the evolutionary development of the urban settlement system from $K = 2$ in 1953 to $K = 3$ in 1985.

There has been no optimisation of the transport structure. The latter modification is still in place today. The forecast of the evolution of the settlement system shows that the modification of $K = 3$ will continue in 2030 even in the case of unification of the Republic of Korea with the DPRK.

**List of references**

1.    Abragimovich E.G., Naimark N.I. Method for quantitative analysis and forecasting the spatial structure of urban settlements network at the regional level. Ser. geogr. 1978. №2. C. 70 - 78.

2.    Alekseev P.V., Panin A.V. Philosophy: Textbook. Moscow: Prospect, 2007. 592 c.

3.    Baransky N.N. Methodology of Teaching Economic Geography. Moscow: Prosveshcheniye, 1990. 303 c.

4.    Baransky N.N. On the economic and geographical study of cities // Voprosy geografii. 1946. №2. C. 19 - 62.

5.    Bezrukov L.A. Transport-geographical continentality of Russia: concept, assessment, dynamics // Izvestia RAN. Ser. geogr. 2004. №5. C. 15 - 25.

6.    Bugromenko V.N. Transport in territorial systems. Moscow: Nauka, 1987. 112 c.

7.    Buchkin A.A. Urbanization, Migration and the Problem of Employment // Social Evolution of Modern South Korea. Moscow: Nauka, 1987. C. 50 -75.

8.    Vazhenin A.A. Hierarchy of central places and regularities in the development of settlement systems // Izvestiya RAN. Ser. geogr. 2002. №5. C. 64 - 71.

9.    Vazhenin A.A. Stability of distribution of urban settlements in settlement systems // Izvestia RAN. Ser. geogr. 1999. №1. C. 55 - 59.

10.    Vazhenin A.A. Evolutionary processes in settlement systems. Ekaterinburg: UrO RAN, 1997. -60 c.

11.    Vazhenin A.A. Evolution of spatial settlement structures: changing patterns // Izvestia RAN. Ser. geogr. 2006. №3. C. 29 - 38.

12.    Vakhnenko R.V. Transport and mobility of the population of the south of the Far East. Vladivostok: FEB RAS Publishing House, 1990. 139 c.

13.    Vlasov P., Viktorov S. "Tiger Caught in the Storm // Expert. 1998. №3. C. 21 - 24.

14.    Garner J. Models of Urban Geography and Placement of Settlements // Models in Geography. Moscow: Progress, 1971. C. 29 - 86.

15.    Gladkov I. Seoul in the focus of the country's economic life // Asia and Africa Today. 2002. №11. C.20 - 26.

16.    Glushko A.A., Ryabinina L.I. Territorial structure of the economy of APR countries. Vladivostok: TIDOT, 2004. 178 c.

17.    Davidovich V.G. Quantitative regularities of interconnected settlement in urban

agglomerations // District planning and settlement. - Kyiv, 1968. C. 34 - 57.

18.  Dmitriev A.V. Where the Soviet Man Lives. Moscow: Mysl, 1988. 223 c.

19.  Domański Ryszard. Economic Geography: A Dynamic Aspect. Moscow: The New Chronograph, 2010. 376 c.

20.  Zayochkonskaya J. Methodology and Methods of Studying Migration Processes. Moscow: Centre for Migration Studies, 2007. 370 c.

21.  Zorin I.V., Kantsebovskaya I.V. Some methods for measuring the level of urbanisation // Problems of Modern Urbanisation. Moscow: Statistics, 1972. C. 191 - 203.

22.  Jelonek A. Population Formation, Urbanisation Processes and Changing Demographic Zoning in Poland // Problems of Urbanisation and Resettlement. Moscow: Mysl, 1976. C. 63 - 69.

23.  Erczyński M. Issues of development and territorial and functional organisation of the urban system in Poland // Problems of urbanisation and resettlement. Moscow: Mysl, 1976. C. 98 - 117.

24.  Konstantinov O.A. Types of urbanization in the USSR // Izvestia of the USSR Academy of Sciences. Ser. geogr. 1976. № 5. C. 72 - 83.

25.  Korea: a handbook. Seoul: Korea Overseas Information Service, 1993. 676 c.

26.  Kostinsky G.D. Republic of Korea // Geography at School. 2002. №5. C. 34 - 45.

27.  Krylov P.M. Typology of modern regional transport systems in Russia // Izvestia RAN. Ser. geogr. 2007. №4. C. 66 - 75.

28.  Kuznetsova N. V., Fokin N. I. The Potential of Economic Development of the Republic of Korea. Vladivostok: Far East University Press, 2003. 300 c.

29.  Kümmel T. The Stadial Concept of Urbanisation: Methodology and Methods of Analysis. MOSCOW: IGAN, 1987. C. 82 - 100.

30.  Lankov A. Seoul: city-state. [Electronic resource] - Available from URL: http://world.lib.ru/kZkim_o_i/a2-1.shtml [accessed 14.04.2008].

31.  Lappo G.M. Geography of Cities: Textbook for Geography Departments of Higher Education Institutions. MOSCOW: VLADOS, 1997. 480 c.

32.  Lösch A. The spatial organisation of the economy. Moscow: Nauka, 2007. 663 c.

33.  Maergoyz I.M. Some Issues of Studying the Economic and Geographic Position of Cities in the USSR // Proceedings of the II Interdisciplinary Council on Population Geography. MOSCOW: MFGO, 1968. Vol. 1. C. 92 - 98.

34. Mironenko N.S. Country Studies: Theory and Methods: Textbook for Higher Education Institutions. Moscow: Aspect-Press, 2001. 268 c.

35. Nefedova T.G., Treyvish A.I. The theory of "differential urbanisation" and the hierarchy of cities in Russia at the turn of the 21st century // Problems of urbanisation and settlement at the turn of the century. Smolensk: Oikumen, 2002. C. 71 - 87.

36. Ozerova G.N., Pokshishevsky V.V. Geography of the World Urbanisation Process. Moscow: Prosveshcheniye, 1981. 189 c.

37. Percik E.N. Geo-urbanism: textbook for students of higher education institutions. Moscow: Academy of Sciences, 2009. 432 c.

38. Petrov N.V. Spatial and temporal analysis of settlement systems: methodological aspects (on the example of Moscow) // Methods for studying settlement. MOSCOW: IGAN, 1987. C. 55 - 68.

39. Pivovarov Y.L. Fundamentals of Geo-Urbanistics: Urbanisation and Urban Systems: Textbook for Higher Education Institutions. MOSCOW: VLADOS, 1999. 232 c.

40. Pityurenko E.I. Systems of Settlement and Territorial Organization of National Economy. Kyiv: Naukova Dumka, 1983. 206 c.

41. Polyan P.M. Methodology of identifying and analysing the basic framework of settlement. MOSCOW: INSTITUTE OF THE USSR ACADEMY OF SCIENCES, 1988. 220 c.

42. Polyan P.M. Methods for estimating the territorial concentration of population // Methods for studying settlement. MOSCOW: IGAN, 1987, PP. 164 - 176.

43. Polyan P.M. Urbanisation and methods for its assessment // Izv. of the Academy of Sciences of the USSR. Ser. geogr. 1980. № 5. C. 63 - 77.

44. Popov R.A. Quantitative characteristics of the urbanization of Russia's regions in the second half of the twentieth century. Ser. geogr. 2002. № 1. C. 49 - 56.

45. Semina I.A. Methodology of studying regional transport infrastructure (by the example of Mordovia) // Izv. Ser. geogr. 2009. №1. C. 48 - 56.

46. Simagin Y.A. Territorial organization of population: textbook. Moscow: Dashkov & Co. 236 c.

47. Taborisskaya I.M. Pendulum migration of population: theory, methodology, practice. Moscow: Statistics, 1979. 177 c.

48. Tarhov S A. Evolutionary morphology of transport networks: methods of topological regularities analysis. MOSCOW: INSTITUTE OF THE USSR ACADEMY OF

SCIENCES, 1989. 221 c.

49.    Topchiev A.K. Formalized analysis and assessment of transport and geographical position of cities // Bulletin of Moscow State University. Ser. geogr. 1974. №4. C. 47 - 54.

50.    Torkunov A.V. History of Korea. MOSCOW: RSE, 2003. 637 c.

51.    Haggett, P. Geography: a synthesis of modern knowledge. Moscow: Progress, 1979. - 684 c.

52.    Khanin S.E. Modelling territorial systems // Methods for studying settlement. MOSCOW: IG AS. 1987. C. 131 - 144.

53.    Kharitonov V.M. Urbanisation in the United States. Moscow: Moscow State University Press, 1983. 200 c.

54.    Hezman S. Peculiarities of urbanisation in Poland in 1960-1970 // Problems of urbanisation and resettlement. Moscow: Mysl, 1976. C. 37 - 53.

55.    Khorev B. et al. Migration Studies: Textbook. Moscow: Moscow State University Press, 1989.

56.    Khudyaev I.A. Evolution of settlement systems: from regularity to singularity [Electronic resource    ]                 - Available from              URL: www.shu.ru/pages/mag/RI_2008_04(19).pdf [Accessed 17.01.2010].

57.    Shuper V.A. Study of socio-geographical space metrics (on the example of the centre of the European part of the RSFSR). Author's abstract for the degree of Candidate of Sciences in Geography. MOSCOW: INSTITUTE OF HISTORY OF THE USSR, 1980. 19 c.

58.    Shuper V.A. Self-organisation of the urban population. MOSCOW: ROU, 1995. 167 c.

59.    Shuper V.A. Stability of the spatial structure of urban settlement systems. Doctoral thesis in Geography. MOSCOW: INSTITUTE OF HISTORY OF THE USSR, 1990. 38 c.

60.    Em P.P. Analysis of transport system development level in the Republic of Korea // Proceedings of XVII International Conference of Students, Postgraduates and Young Scientists "Lomonosov - 2010". [Electronic resource]. Moscow: MAKS Press, 2010. CD-ROM.

61.    Em P. P. Hierarchy of urban settlement structure of the Republic of Korea // Proceedings of the scientific conference of undergraduate and graduate students of FEFU - 2009. Vladivostok: Far Eastern University Publisher, 2009. C. 219 - 223.

62.    Em P.P. Quantification of Urbanization and Pendular Migration in the Republic of

Korea // Studies of Young Geographers: Collection of articles by the winners of the "Geography" section of the XVII International Scientific Conference "Lomonosov". - Moscow: Faculty of Geography, Moscow State University, 2009. C. 120 - 124.

63.    Em P.P. Peculiarities of pendular migration in the Republic of Korea // Geographical study of territorial systems: collection of materials of III All-Russian scientific - practical conference. Perm: Publishing house of Perm State University, 2009. C. P. 235 - 238.

64.    Em P.P., Ryabinina L.I. Assessment of transport and geographical potential of the largest cities of the Republic of Korea // Geography and contemporary problems of natural science cognition: materials of All-Russian scientific and practical conference. Ekaterinburg: Ural State Pedagogical University Publisher, 2009. C. 127 - 130.

65.    Christaller W. Central place in Southern Germany. Englewood Cliffs. N.J., 1966.

66.    Korean Statistical Information Service [Electronic resource] - Available from URL: http://www.kosis.kr [Accessed 20.02.2010].

67.    Korea Statistical Yearbook. Seoul: National Statistical Office of the Republic of Korea, Vol. 56, 2009.

68.    National Statistical Office of the Republic of Korea [Electronic resource] - Available from URL: http://www.nso.go.kr [Accessed 20.02.2010].

69.    The official web-page of Seoul Metropolitan City [Electronic resource] - Available from URL: http://english.seoul.go.kr/ [Accessed 25.06.2009].

70.    The World Factbook [Electronic resource] - Available from URL: https://www.cia.gov/library/publications/the-world-factbook/geos/hk.html        [Date accessed 17.03.2010].

71.    The World Urbanization Prospects: Revision of 2009 [Electronic resource] - Available from URL: http://esa.un.org/unup/ [Accessed 15.12.2009].

REFORMENT

Table 1

Example of determining distances and angles for plotting points on a settlement system map in 2009. (compiled by the author on the basis of calculated data)

| City number | Distance from the beginning of the beam to the borders of the country, km | Distance from start of beam to point, km | Proportion of distance from beam start to point to distance from beam start to country borders, % | Distance plotted on ellipsoid with R=60 mm, mm | Initial degree of sedimentation | Degree used |
|---|---|---|---|---|---|---|
| Seoul | 0 | 0 | 0 | 0 | 0 | 0 |
| Phaju | 39,9 | 30,45 | 76,3 | 45,8 | 47 | 94 |
| Goyang | 40,9 | 17,8 | 43,6 | 26 | 198,5 | 397 |
| Gunpo | 44,1 | 26,2 | 59,5 | 35,7 | 194 | 388 |
| Bucheon | 36,7 | 18,9 | 51,4 | 30,8 | 159 | 318 |
| Incheon | 35,7 | 29,4 | 82,4 | 49,4 | 154 | 308 |
| Siheung | 32,5 | 27,3 | 83,9 | 50,3 | 139 | 278 |
| Ansan | 51,4 | 32,5 | 63,3 | 38 | 126 | 252 |
| Hwasong | 56,7 | 43 | 76 | 45,5 | 110 | 220 |
| Euwang | 483 | 24,1 | 5 | 3 | 95 | 190 |
| G Wachhon | 342,3 | 14,7 | 4,3 | 2,6 | 91 | 182 |
| Suwon | 318,1 | 31,5 | 10 | 6 | 87 | 174 |
| Osan | 304,5 | 45,1 | 15 | 8,8 | 81 | 162 |
| Pyeongtaek | 304,5 | 64 | 21 | 12,6 | 81 | 162 |
| Ansong | 310,8 | 64 | 20,6 | 12,3 | 68 | 136 |
| Yongin | 345,4 | 24,2 | 7 | 4 | 91 | 182 |
| Seongnam | 311,8 | 17,8 | 5,8 | 3,5 | 43 | 86 |
| Gwanju | 270,9 | 27,3 | 10,1 | 6 | 38 | 76 |
| Icheon | 270,9 | 49,3 | 18,2 | 10,9 | 38 | 76 |
| Khanam | 194,2 | 19,9 | 10,3 | 6,2 | 3 | 6 |
| Guri | 161,7 | 12,6 | 7,8 | 4,7 | 342 | 324 |
| Namyangju | 161,7 | 18,9 | 11,7 | 7 | 342 | 324 |
| Ejonbu | 49,3 | 17,8 | 36,2 | 21 | 279 | 198 |
| Yanju | 49,3 | 25,2 | 51,1 | 30,6 | 279 | 198 |
| Donduchen | 49,3 | 37,8 | 76,6 | 46 | 279 | 198 |
| Pocheon | 54,6 | 74,5 | 72,4 | 43,5 | 331 | 320 |
| Cheonan | 304,5 | 86,1 | 28,3 | 17 | 81 | 162 |
| Asan | 318,1 | 81,9 | 25,7 | 15,4 | 87 | 174 |
| Seosan | 117,6 | 100,8 | 85,7 | 51,4 | 119 | 238 |
| Boreng | 161,7 | 139,6 | 86,4 | 51,8 | 104 | 208 |
| Conju | 344,4 | 121,8 | 35,6 | 21,2 | 85 | 170 |
| Nonsan | 318,1 | 151,2 | 47,5 | 28,5 | 87 | 174 |
| Gereng | 304,5 | 143,8 | 47,2 | 28,3 | 81 | 162 |
| Taejon | 306,6 | 138,6 | 45,2 | 27,1 | 77 | 154 |
| Cheongju | 318,1 | 195,3 | 61,4 | 36,8 | 87 | 174 |
| Ixan | 342,2 | 180,6 | 52,8 | 31,7 | 91 | 182 |
| Gimche | 331,8 | 196,3 | 59,2 | 35,5 | 93 | 186 |
| Gunsan | 336 | 176,4 | 52,5 | 31,5 | 98 | 196 |
| Chongyib | 331,8 | 221,5 | 66,8 | 40 | 93 | 186 |
| Namwon | 304,5 | 242,5 | 79,6 | 47,8 | 81 | 162 |
| Suncheon | 304,5 | 295 | 96,9 | 58,1 | 81 | 162 |
| Gwangju | 342,2 | 268,8 | 78,5 | 47,1 | 91 | 182 |
| Nachu | 483 | 283,5 | 58,7 | 35,2 | 95 | 190 |

| | | | | | |
|---|---|---|---|---|---|
| Mokpo | 354,9 | 311,8 | 87,8 | 52,7 | 100 | 200 |
| Cheongnam | 336 | 310,8 | 92,5 | 55,5 | 98 | 196 |
| Esu | 318,1 | 312,9 | 98,3 | 59 | 79 | 158 |
| G vanyang | 306,6 | 298,2 | 97,2 | 58,3 | 77 | 154 |
| Jinju | 310,8 | 280,3 | 90 | 54 | 68 | 136 |
| Sachon | 310,8 | 302,4 | 97,3 | 58,4 | 68 | 136 |
| Thongyeong | 328,6 | 328,6 | 100 | 60 | 66 | 132 |
| Codje | 345,4 | 332,8 | 96,3 | 57,8 | 63 | 124 |
| Masan | 298,2 | 298,2 | 100 | 60 | 61 | 122 |
| Changwon | 319,2 | 300,3 | 94,1 | 56,4 | 59 | 118 |
| Jinhae | 319,2 | 310,8 | 97,4 | 58,4 | 59 | 118 |
| Kyme | 329,7 | 307,6 | 93,3 | 55,9 | 55 | 110 |
| Busan | 329,7 | 31,4 | 100 | 60 | 55 | 110 |
| Yangsan | 330,7 | 307,6 | 93 | 55,8 | 52 | 104 |
| Miryang | 334,9 | 279,3 | 83,4 | 50 | 54 | 108 |
| G wangmeng | 317,4 | 286,2 | 90,2 | 54,1 | 48 | 96 |
| Ulsan | 312,9 | 302,4 | 96,6 | 57,9 | 46 | 92 |
| Gyeongju | 311,8 | 276,1 | 90 | 54 | 43 | 86 |
| Pohang | 270,9 | 260,8 | 99 | 59,4 | 38 | 76 |
| Youngchon | 311,8 | 245,7 | 78,8 | 47,3 | 45 | 90 |
| Gyeongsan | 321,3 | 248,8 | 77,4 | 46,4 | 50 | 100 |
| Tag | 330,7 | 236,2 | 71,4 | 42,8 | 52 | 104 |
| Gumi | 330,7 | 201,6 | 60,6 | 36,5 | 52 | 104 |
| Gimcheon | 321,3 | 191,1 | 59,5 | 35,7 | 57 | 114 |
| Sangju | 321,3 | 162,7 | 50,6 | 30,3 | 50 | 100 |
| Mungeng | 311,8 | 149,1 | 47,8 | 28,7 | 45 | 90 |
| Andong | 263,5 | 189 | 71,7 | 43 | 34 | 68 |
| Yongju | 252 | 187,9 | 74,6 | 44,7 | 30 | 60 |
| Taebek | 220,5 | 181,6 | 82,4 | 49,4 | 14 | 28 |
| Jaechon | 238,3 | 115,5 | 48,4 | 29 | 24 | 48 |
| Wonju | 223,6 | 87,1 | 38,9 | 23,3 | 16 | 32 |
| Samchok | 194,2 | 192,1 | 98,9 | 59,3 | 3 | 6 |
| Donghae | 193,2 | 186,9 | 96,7 | 58 | 0 | 0 |
| Sokcho | 173,2 | 165,9 | 95,7 | 57,5 | 171 | 342 |
| Chuncheon | 102,9 | 74,5 | 72,4 | 43,5 | 331 | 302 |
| Jeju | 483 | 456,7 | 94,6 | 56,8 | 95 | 190 |
| Sogwipho | 477,7 | 477,7 | 100 | 60 | 94 | 188 |
| Anjang | 354,9 | 18,9 | 5 | 3 | 100 | 200 |
| Cheongju | 328,6 | 111,3 | 33,8 | 20,3 | 66 | 132 |
| Chungju | 270,9 | 103,9 | 38,4 | 23 | 38 | 76 |
| Gangnam | 173,2 | 165,9 | 95,7 | 57,5 | 171 | 342 |

Example of determining distances and angles north of the reference line for

the mapping of the settlement system in 2030. (forecast)

(compiled by the author on the basis of calculated data)

| City number | Distance from the beginning of the beam to the country border, km | Distance from start of beam to point, km | Proportion of distance from beam start to point to distance from beam start to country borders, % | Distance plotted on ellipsoid with R=60 mm, mm | Initial degree of sedimentation | Degree used |
|---|---|---|---|---|---|---|
| Gunpo | 51,4 | 25,2 | 49 | 29,4 | 14 | 28 |
| Goyang | 199,5 | 16,8 | 8,4 | 5 | 35 | 70 |
| Haeju | 196,3 | 121,8 | 62 | 37,2 | 29 | 58 |
| Sarivon | 226,8 | 151,2 | 66,7 | 39,9 | 43 | 86 |
| Nampho | 226,8 | 194,2 | 85,6 | 51,5 | 43 | 86 |
| Kaesong | 225,7 | 60,9 | 27 | 16,2 | 48 | 96 |
| Phaju | 225,7 | 30,4 | 13,5 | 8,1 | 48 | 96 |
| Sanyju | 362,2 | 358 | 98,8 | 59,3 | 52 | 104 |
| Pyongyang | 367,5 | 195,3 | 53 | 31,9 | 56 | 112 |
| Cheongju | 367,5 | 284,5 | 77,4 | 46,5 | 56 | 112 |
| Goosong | 392,7 | 310,8 | 79,1 | 47,5 | 61 | 122 |
| Pyeongsong | 373,8 | 213,1 | 57 | 34,2 | 63 | 126 |
| Anju | 373,8 | 256,2 | 68,5 | 41,1 | 63 | 126 |
| Sunjong | 373,8 | 226,8 | 60,7 | 36,4 | 65 | 130 |
| Togjeong | 384,3 | 253 | 65,8 | 39,5 | 75 | 150 |
| Heucheon | 387,4 | 298,2 | 76,9 | 46,2 | 77 | 154 |
| Gange | 434,7 | 382,2 | 87,9 | 52,7 | 85 | 170 |
| Yanju | 436,8 | 26,2 | 6 | 4 | 99 | 198 |
| Donduchen | 436,8 | 37,8 | 8,6 | 5,2 | 99 | 198 |
| Ejonbu | 436,8 | 18,9 | 4,3 | 2,6 | 99 | 198 |
| Wonsan | 436,8 | 191,1 | 43,7 | 26 | 99 | 198 |
| Hamhyung | 436,8 | 265,6 | 60,8 | 36,5 | 99 | 198 |
| Simpho | 582,7 | 295 | 68,6 | 41,2 | 109 | 218 |
| Pocheon | 636,3 | 40,9 | 6,4 | 4 | 116 | 232 |
| Dapcheon | 636,3 | 360,1 | 56,6 | 33,9 | 116 | 232 |
| Chongjin | 636,3 | 529,2 | 83,2 | 49,9 | 116 | 232 |
| Nason | 591,1 | 591,1 | 100 | 60 | 118 | 236 |
| Namyangju | 156,4 | 18,9 | 12 | 7 | 151 | 302 |
| Chuncheon | 156,4 | 75,6 | 48 | 29 | 151 | 302 |
| Wonsan | 156,4 | 156,4 | 100 | 60 | 151 | 302 |
| Guri | 156,4 | 11,5 | 7,4 | 4,4 | 157 | 317 |
| Sokcho | 173,2 | 165,9 | 95,7 | 57,5 | 171 | 342 |

Forecast of the size of the largest cities in the Republic of Korea in 2030.

(compiled by the author on the basis of calculated data)

| № | City | Population size in 2004 | Population size in 2009 | Growth rate (r) | Ert | Population forecast in 2030 |
|---|---|---|---|---|---|---|
| 1 | Seoul | 23431 | 23900 | 0,11 | 1,046 | 24724 |
| 2 | Busan | 3865 | 3615 | -0,64 | 0,77 | 2989 |
| 3 | Tag | 2501 | 2513 | 0,01 | 1,05 | 2611 |
| 4 | Taejon | 1323 | 1488 | 0,12 | 1,64 | 2171 |
| 5 | Gwangju | 1326 | 1423 | 0,07 | 1,34 | 1773 |
| 6 | Ulsan | 1013 | 1113 | 0,09 | 1,48 | 1499 |
| 7 | Youngjin | 302 | 814 | 1,69 | 1,06 | 322 |
| 8 | Ansan | 651 | 735 | 0,13 | 1,67 | 1086 |
| 9 | Cheongju | 545 | 638 | 0,17 | 1,96 | 1073 |
| 10 | Anjang | 593 | 631 | 0,06 | 1,28 | 764 |
| 11 | Cheongju | 590 | 627 | 0,06 | 1,28 | 756 |
| 12 | Cheonan | 473 | 541 | 0,14 | 1,77 | 837 |
| 13 | Changwon | 508 | 510 | 0,003 | 1,01 | 516 |
| 14 | Pohang | 512 | 509 | -0,005 | 0,97 | 500 |
| 15 | Gime | 407 | 475 | 0,17 | 1,94 | 790 |
| 16 | Namyanju | 429 | 498 | 0,16 | 1,89 | 812 |
| 17 | Ejonbu | 429 | 498 | 0,12 | 1,65 | 625 |
| 18 | Masan | 377 | 425 | -0,02 | 0,91 | 393 |
| 19 | Jeju | 316 | 418 | 0,14 | 1,75 | 643 |
| 20 | Siheung | 374 | 405 | 0,08 | 1,39 | 519 |
| 21 | Gumi | 365 | 396 | 0,08 | 1,4 | 511 |

# I want morebooks!

Buy your books fast and straightforward online - at one of world's fastest growing online book stores! Environmentally sound due to Print-on-Demand technologies.

Buy your books online at
**www.morebooks.shop**

Kaufen Sie Ihre Bücher schnell und unkompliziert online – auf einer der am schnellsten wachsenden Buchhandelsplattformen weltweit! Dank Print-On-Demand umwelt- und ressourcenschonend produziert.

Bücher schneller online kaufen
**www.morebooks.shop**

Printed by Books on Demand GmbH, Norderstedt / Germany